Fundamentals of Brooks–Iyengar Distributed Sensing Algorithm

Pawel Sniatala • M. Hadi Amini
Kianoosh G. Boroojeni

Fundamentals of Brooks–Iyengar Distributed Sensing Algorithm

Trends, Advances, and Future Prospects

 Springer

Pawel Sniatala
Department of Computing Science
Poznań University of Technology
Poznań, Poland

Kianoosh G. Boroojeni
School of Computing
and Information Sciences
Florida International University
Miami, FL, USA

M. Hadi Amini
School of Computing
and Information Sciences
Florida International University
Miami, USA

ISBN 978-3-030-33134-4   ISBN 978-3-030-33132-0   (eBook)
https://doi.org/10.1007/978-3-030-33132-0

This Springer imprint is published by the registered company Springer Nature Switzerland AG.
The registered company address is: Gewerbestrasse 11, 6330 Cham, Switzerland

Endorsement

"The book presents emerging applications and influence of Brooks–Iyengar distributed sensing algorithm. It will help audience to better understand the theory of distributed sensor networks and its applications." Prof. Jacek Błażwicz, IEEE Fellow, Director, Institute of Computing Science and Institute of Bioorganic Chemistry, Poznan University of Technology, Poland.

"I highly recommend this book to scientists, researchers, engineers, and practitioners who are interested in distributed sensor networks. This book thoroughly introduces strong theoretical techniques for real-world applications in the context of distributed sensing." Prof. Poki Chen, Dean, College of Applied Sciences; Professor, Dept. of Electronic and Computer Engineering; National Taiwan University of Science and Technology.

"This book, *Fundamentals of Brooks–Iyengar Distributed Sensing Algorithm*, provides a very thorough treatment of the famous Brooks–Iyengar algorithm for distributed sensor networks and related methods. Distributed sensor fusion techniques are very important for many real-time critical applications in unstructured environment. This timely book introduces a novel algorithmic computational platform underlying the paradigm of the Brooks–Iyengar algorithmic structure to efficiently solve sensor fusion problems in unpredictable and Byzantine circumstances. Furthermore, performance bounds on the Brooks–Iyengar algorithm have been characterized elegantly by discrete mathematical theory, which provides a foundation for a large number of real-time applications." Prof. Sartaj Sahni, Distinguished Professor, Computer and Information Sciences and Engineering, University of Florida, USA; IEEE Fellow, ACM Fellow, AAAS Fellow, Minnesota Supercomputer Institute Fellow.

Theme of the Book

Brooks–Iyengar Distributed Sensing Algorithm has made quite a lot of impact since its initial publication in 1996. It is classic and insightful even if one reads it today. We see the technique has been applied in many domains such as software reliability, distributed systems, and OS development. We would expect Brooks–Iyengar algorithm to continue to be used where fault-tolerant solutions are needed in redundancy system scenarios.

This book highlights many of the recent applications of Brooks–Iyengar algorithm in the context of foundational ideas and applications. "The contribution of the Brooks–Iyengar Distributed Computational Sensing work has enhanced new real-time features by adding fault-tolerant capabilities for many applications." Results of the works produced by Dr. Iyengar, his students, and colleagues have been compiled in this book with permission.

Preface

This book provides a comprehensive analysis of Brooks–Iyengar Distributed Sensing Algorithm, which brings together the power of Byzantine Agreement and sensor fusion in building a fault-tolerant distributed sensor network. The authors analyze its long-term impacts, advances, and future prospects. The book starts by discussing the Brooks–Iyengar algorithm, which has made significant impact since its initial publication in 1996. The authors show how the technique has been applied in many domains such as software reliability, distributed systems, and OS development. The book exemplifies how the algorithm has enhanced new real-time features by adding fault-tolerant capabilities for many applications. The authors posit that the Brooks–Iyengar algorithm will continue to be used where fault-tolerant solutions are needed in redundancy system scenarios.

This book:

- Provides a comprehensive investigation of Brooks–Iyengar algorithm and the recently developed methods based on this algorithm.
- Specifies the importance and position of theoretical methods in dealing with real-world problems in the context of distributed sensing.
- Presents basics and mathematical foundations needed to analyze Brooks–Iyengar algorithm.

Poznań, Poland   Pawel Sniatala
Miami, FL, USA   M. Hadi Amini
Miami, FL, USA   Kianoosh G. Boroojeni

Acknowledgments

The authors would like to thank their colleagues and friends for their continuous support.

Contents

Contents

About the Authors

Pawel Sniatala received his M.S. degree in Telecommunication, M.S. degree in Computer Science, and Ph.D. degree in Microelectronics all from Poznan University of Technology, Poland. He received his habilitation degree (D.Sc.) in electronics in 2016. From 1998 to 2002 he was with the Department of Computer Engineering at the Rochester Institute of Technology (USA). He returned to Poznan University of Technology to take a position in the Faculty of Computing, where currently he is the Vice Dean for Industrial Cooperation. He also graduated the international MBA study in a joint program of Georgia State University and Poznan University of Economics. His area of interests focuses on VLSI circuits for digital and mixed analog-digital signal processing systems: UltraLow Power ASIC design, implantable IC, signal/image processing hardware-software codesign, hardware accelerators. However, following his computing science background, he is also involved in research projects related to eHealth area. He was involved in several industrial projects, e.g., control and monitoring systems for gas mine systems, control systems for water treatment plant, and teletechniques systems for airport. He has served in several international Ph.D. and M.S. committees (Portugal, USA). He has presented invited courses/lectures in several universities outside Poland: USA, Portugal, Peru, Taiwan. He is an author and coauthor of 90 papers including a monograph: P. Sniatala, CMOS Current Mode Modulators, Poznan Monographs in Computing and Its Applications, Poznan 2016.

M. Hadi Amini received his Ph.D. and M.Sc. from Carnegie Mellon University in 2019 and 2015 respectively. He also holds a doctoral degree in Computer Science and Technology. Prior to that, he received M.Sc. from Tarbiat Modares University in 2013, and the B.Sc. from Sharif University of Technology in 2011. He is currently an Assistant Professor at School of Computing and Information Sciences, College of Engineering and Computing at Florida International University (FIU). He is also the founding director of Sustainability, Optimization, and Learning for InterDependent networks laboratory (solid lab). His research interests include distributed machine learning and optimization algorithms, distributed intelligence, sensor networks, interdependent networks, and cyberphysical resilience. Application domains include energy systems, healthcare, device-free human sensing, and transportation networks.

Prof. Amini is a life member of IEEE-Eta Kappa Nu (IEEE-HKN), the honor society of IEEE. He organized a panel on distributed learning and novel artificial intelligence algorithms, and their application to healthcare, robotics, energy cybersecurity, distributed sensing, and policy issues in 2019 workshop on artificial intelligence at FIU. He also served as President of Carnegie Mellon University Energy Science and Innovation Club; as technical program committee of several IEEE and ACM conferences; and as the lead editor for a book series on "Sustainable Interdependent Networks" since 2017, as well as. He has published more than 80 refereed journal and conference papers, and book chapters. He has co-authored two books, and edited three books on various aspects of optimization and machine learning for interdependent networks. He is the recipient of the best paper award of "IEEE Conference on Computational Science & Computational Intelligence" in 2019, best reviewer award from four IEEE Transactions, the best journal paper award in "Journal of Modern Power Systems and Clean Energy", and the dean's honorary award from the President of Sharif University of Technology. (homepage: www.hadiamini.com; lab website: www.solidlab.network).

Kianoosh G. Boroojeni received his PhD in computer science from Florida International University. He received his B.Sc degree from the University of Tehran in 2012. His research interests include smart grids and cybersecurity. He is the author/coauthor of a number of books published by MIT Press and Springer and various peer-reviewed journal publications and conference proceedings. He is currently a faculty at Florida International University.

Acronyms

The following are the abbreviations of the terms used in this book.

ABA   Approximate Byzantine Agreement
BGP   Byzantine Generals Problem
BVC   Byzantine Vector Consensus
DARPA   Defense Advanced Research Projects Agency
FCA   Fast Convergence Algorithm
MRI   Multi-Resolution Integration
PE   Processing Element
WRD   Weighted Region Diagram

The original version of this book was revised: The affiliation of the author M. Hadi Amini
has been corrected. The correction to this book is available at
https://doi.org/10.1007/978-3-030-33132-0_11

Part I
Introduction

Chapter 1
Introduction to Sensor Networks

The evolution of sensory devices and the advancements in wireless communication and digital electronics, brought about a revolution in the way sensor nodes are designed and utilized. The communication mechanism has also seen a complete makeover. Modern sensor networks involve the deployment of multiple miniature sensors across the area of interest wherein sensory data is desired. These miniature devices are specialized for certain purposes and usually possess minimal processing and computing capabilities. With advancements in the field of distributed computing, people started designing distributed sensor networks that would involve multiple such sensors communicating, sharing, and processing information collected by them for a specific goal. A compounded problem with sensors is the inaccuracy and lack of precision in the values collected which could lead to faulty processing. Lots of research in the 1990s reveal that sensor fusion is a powerful method that can be used to mask the failures and minimize the effects of such faulty data. This chapter would try to highlight the applicability of the seminal Brooks–Iyengar hybrid algorithm on distributed sensor networks bringing together the power of Byzantine agreement and sensor fusion in building a fault tolerant distributed sensor network.

1.1 General Description

It has been observed that when sensory devices are brought together to work in unison, the results obtained can be more beneficial. Thus, sensors are clustered to forming networks called sensor networks. Sensor networks has been a subject of great interest since 2000 and the most important reason for this has been the major advances in micro-electro-mechanical systems (MEMS) technology, wireless communications, and digital electronics. Better understanding of semiconductor physics which brought about a revolution in the fabrication and deployment of

© Springer Nature Switzerland AG 2020
P. Sniatala et al., *Fundamentals of Brooks–Iyengar Distributed Sensing Algorithm*,
https://doi.org/10.1007/978-3-030-33132-0_1

low-cost and low-power chips benefitted the creation of sensor nodes. With the processor on chip idea flourishing, incorporating processing elements on these nodes became possible and also made up for more advanced ways of communicating wirelessly in a network and thus brought about the advent of wireless sensor network (WSN). Over the years, WSNs have become more robust. There have been various communication protocols that have been defined to enable the transfer of information, to perceive the environment and make decisions. In most cases, a sensor node constitutes sensing units, low-cost processor, communication module, and a battery [1].

Sensors in general sense or read a physical variable and provide a numerical output which could be inaccurate or get corrupted by the addition of noise. This could lead to the presence of a faulty set of data in the data aggregated at the processing unit. Hence, there is a need for proper calibration and fault tolerant algorithms that could overcome the effects of these faulty data.

1.2   Wireless Sensor Networks

Sensors that were wired for many decades started communicating with no wire through existing protocols or by formulating new ones. This brought about a radical change in how sensors were used. It opened new applications and also made processing possible through the concepts of distributed computing. It also promoted the sharing of data and resources through sensor nodes which are configured in the form of networks. This led to the formation of wireless sensor networks (WSN).

A wireless sensor network (WSN) can be defined as a wireless network of sensors that are spatially distributed and have a specific function in the environment of deployment. A WSN system that have been deployed in the environment are expected to communicate wither continuously or in bursts with the wired or distributed nodes where data aggregation and fusion occurs before the processing. WSNs have many advantages over the traditional sensors in terms of size, the ease of deployment, possibility of collaborative sensing, lower maintenance costs, and self-organizing capabilities to name a few.

Though WSN offers a lot on the plate, there are some issues that have to be addressed. Some of the issues include the possibility of transmission of inaccurate or faulty data. Researchers have come up with multiple algorithms to address the fault tolerance. The further chapters in this book define and discuss in more detail about the Brooks–Iyengar hybrid algorithm that could be used for fault tolerance by the techniques of sensor fusion described in more detail in the future sections of the book.

1.3 Distributed Sensor Networks

Recent years have seen an increasing interest to wars highly advanced systems for surveillance, automation, tracking, etc. and the rapid growth of IoT devices had made this more pronounced. Further, there is a need for efficient algorithms that are tailored for distributed computing using the information from distributed sensor networks for smart cities infrastructure, energy systems, and IoT applications [2–5]. All this brought about the need for the design of spatially distributed systems that was based on the integration of solutions obtained from solving of multiple sub problems and fusing the outcome and thus creation of a distributed sensor network (DSN). An example structure of a DSN is depicted in Fig. 1.1. DSNs constitute of sensors that are spatially distributed in order to detect and measure some phenomena based on its varying parameters as depicted by the sensors. The information collected from these sensor nodes are sent to a central location for processing wherein the collected data is aggregated and fused. As explained earlier, the data transmitted by the sensors may be faulty or inaccurate. To overcome these faulty values, various methods of fault tolerance have been studied and researched. The upcoming chapters of this book would describe the various algorithms proposed for fault tolerance and the throw specific light on the usage of the Brooks–Iyengar hybrid algorithm in this process.

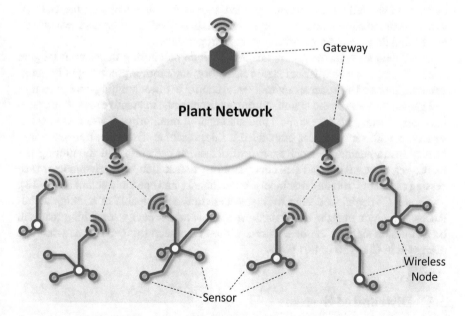

Fig. 1.1 An example structure of a distributed sensor networks (DSN)

1.4 Sensor Networks Applications

There are numerous applications wherein sensors have been successfully deployed. Sensor networks in general may consist of many different types of sensory nodes like the seismic sensors, low sampling rate magnetic sensors, thermal sensors, visual sensors, infrared sensors, acoustic sensors, and radar to name a few. These sensor networks are either deployed alone or as a cluster of multi-sensor nodes having the capability of monitoring a wide variety of ambient conditions that would mostly include sensing of temperature, humidity, vehicular movement, pressure, soil features, noise levels, etc. The sensors could also be used to check for the presence or absence of certain kinds of objects, mechanical stress levels on attached objects, real-time continuous sensing, event detection, location sensing, and even local control of actuators.

The introduction of the IoT technology and the various smart environments open new areas of new applications that arise for these sensors and sensor networks. There are various ways of categorizing the sensor-based applications. We categorize the applications into military, environment, health, home, and other commercial areas. This section gives a brief about some of them. More detailed explanation of example applications would be available in the future sections of the book.

The hospital setup is the easiest place to experience the way in which a sensor-based system can be fully deployed in continuous patient monitoring of the parameters as well as in sending out alert signals and in scheduling the tasks of various hospital staff. Computerized systems as described in [6] have shown proven results that they can help minimize adverse drug events.

Other modern commercial applications include monitoring of material fatigue; building virtual keyboards; managing inventory; monitoring product quality; constructing smart office spaces; environmental control in office buildings; robot control and guidance in automatic manufacturing environments; interactive toys; interactive museums; factory process control and automation; monitoring disaster area; smart structures with sensor nodes embedded inside; machine diagnosis; transportation; factory instrumentation; local control of actuators; detecting and monitoring car thefts; vehicle tracking and detection; and instrumentation of semiconductor processing chambers, rotating machinery, wind tunnels, and anechoic chambers [7–18].

Figure 1.2 depicts a design for a smart environment capable of tracking ambulances in a city using the distributed sensor networks strategy something that will be a common sight in the near future. More about such futuristic ideas would be discussed in Chap. 10 of this book.

1.5 Distributed Systems

The word distributed in terms such as "distributed system," "distributed programming," and "distributed algorithm" originally referred to computer networks where individual computers were physically distributed within some geographical area

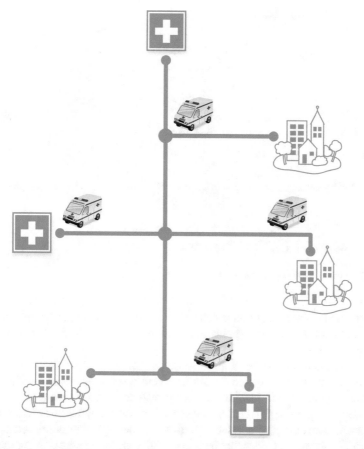

Fig. 1.2 Design of smart environment capable of tracking ambulances in a city using distributed sensor network

[19]. With advances in technology, these words have gathered newer meanings and a broader scope. The shift towards building smart environments and IoT devices that enable human computer interaction and the possibility of even smartphones becoming sensing devices, the term distributed computing joins hands with ubiquitous computing, cyber-physical systems, etc. The growth of machine learning and deep learning along with the fame harnessed by artificial intelligence, distributed systems have advanced to the next level and transcended barriers.

Building a reliable system of many individual systems that would work in unison towards a certain goal is a challenge even to this date. There is a need for an agreement among the devices and consensus on the data being shared and used for attaining the overall goal. There are various real-life applications that demand for this congregated approach and have been blossoming till today's date from the early 1990s. Modern voting systems, blockchains for crypto-currencies, multi-sensor data

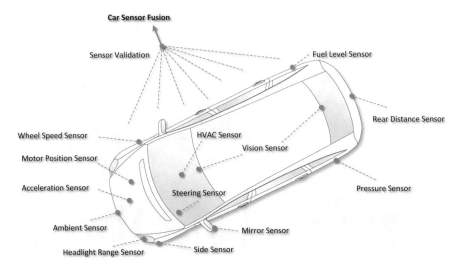

Fig. 1.3 A car as a multi-sensor system

gathering and fusion, synchronized clocks, smart cities, autonomous navigation, etc. is a small chunk of the many applications. An example of a multi-sensor system can be a car (Fig. 1.3).

The agreement or consensus problem hopes for the various agents involved in the network or communication stream to accept a single value or a set of values. The nodes/agents would have to be highly resilient and tolerant to any kinds of faulty data being either injected or transferred due to the failure of some nodes. In such cases, converging to a value that all the nodes agree to is important. There are various models and ideas that help in attaining this agreement. In the literature, multi-agent systems [20] and other distributed algorithms [21] have been deployed for agent-based decision making in complex networks. A commonly used approach in this sense is the majority voting/value algorithm which is the basis on which the Byzantine agreement algorithm and the concepts of Brooks–Iyengar hybrid algorithm are defined. More about these algorithms, their working and practical implications are discussed in the future chapters.

Protocols that solve consensus problems are designed to deal with limited numbers of faulty processes. These protocols must satisfy several requirements to be useful. For instance, a trivial protocol could have all processes output binary value 1. This is not useful and thus the requirement is modified such that the output must somehow depend on the input. That is, the output value of a consensus protocol must be the input value of some process. Another requirement is that a process may decide upon and output a value only once and this decision is irrevocable. A process is called correct in an execution if it does not experience a failure. A consensus protocol tolerating halting failures must satisfy the following properties [22]:

- *Termination:* Every correct process decides some value.
- *Validity:* If all processes propose the same value v, then all correct processes decide v.
- *Integrity:* Every correct process decides at most one value, and if it decides some value v, then some process must have proposed v.
- *Agreement:* Every correct process must agree on the same value.

A protocol that can correctly guarantee consensus among N processes of which at most t fail is said to be t-resilient. In evaluating the performance of consensus protocols two factors of interest are running time and message complexity. Running time is given in Big O notation in the number of rounds of message exchange as a function of some input parameters (typically the number of processes and/or the size of the input domain). Message complexity refers to the amount of message traffic that is generated by the protocol. Other factors may include memory usage and the size of messages.

1.6 Sensor Fusion

A problem with sensors that has always been bothering researchers is that of inaccuracy and lack of precision in the values collected by them which could lead to faulty processing. Lots of research in the 1990s reveal that sensor fusion is a powerful method that can be used to mask and minimize the effects of such faulty data.

Sensor fusion is a term that is associated with combining of sensory data or data derived from multiple sources such that the resulting information has less uncertainty than would be possible when these sources were used individually. The term uncertainty reduction in this case can mean more accurate, more complete, or more dependable, or refer to the result of an emerging view, such as stereoscopic vision (calculation of depth information by combining two-dimensional images from two cameras at slightly different viewpoints) [23, 24]. The data sources for a fusion process are not specified to originate from identical sensors. One can distinguish direct fusion, indirect fusion, and fusion of the outputs of the former two. Direct fusion is the fusion of sensor data from a set of heterogeneous or homogeneous sensors, soft sensors, and history values of sensor data, while indirect fusion uses information sources like a priori knowledge about the environment and human input. One of a possible illustration of sensor fusion application can be a car (Fig. 1.4). We can find there a lot sensors. Some of them are used to provide the same information (e.g., distance measure) and sensor fusion allows to increase the probability of the accurate enough result. Some commonly used algorithms are based on probability theories, which combines the values from different sensors into a single value that is more confident than each single sensor reading. Some widely used algorithms are listed in the following subsections.

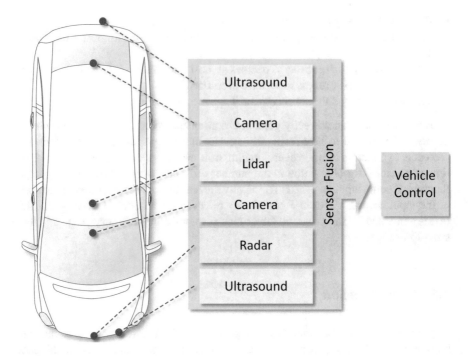

Fig. 1.4 Car's sensors as sources of data for a sensor fusion algorithm

1.6.1   Bayesian Filter

Bayesian filter refers to an algorithm that is commonly employed in computer science for determining the probabilities of multiple beliefs and aiding an object in identifying its position and orientation. These filters allow for estimation of a previously undetermined probability density function (PDF) using the incoming measurement values from the most recently acquired sensor data recursively over time. It consists of three parts: filtering, smoothing, and prediction.

- Filtering: when we estimate the current value given past and current observations,
- Smoothing: when estimating past values given present and past measures,
- Prediction: when estimating a probable future value given the present and the past measures.

Bayesian filters also utilize a mathematical model depicted in Fig. 1.5 in determining the PDF, when the true state is assumed to be an unobserved Markov process, and the measurements are the observed states of a hidden Markov model (HMM).

Fig. 1.5 A typical model
used in Bayesian Filters

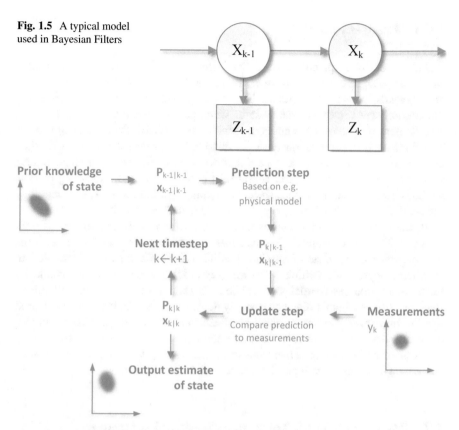

Fig. 1.6 Working of the Kalman filter [25]

1.6.2 Kalman Filter

Kalman filters are the most widely used variant of Bayes filters. Kalman filters' main
advantage is their computational efficiency. We can implement both the prediction
and correction using efficient matrix operations on the mean and covariances as
depicted in Fig. 1.6. This efficiency, however, comes at the cost of restricted repre-
sentational power because Kalman filters can represent only unimodal distributions.
So, Kalman filters are the best if the uncertainty in the state is not too high, which
limits them to location tracking using either accurate sensors or sensors with high
update rates. Despite Kalman filters' restrictive assumptions, practitioners have
applied them with great success to various tracking problems, where the filters yield
efficient, accurate estimates, even for some highly nonlinear systems.

1.6.3   Particle Filter

Particle filters or sequential Monte Carlo (SMC) methods are a set of genetic, Monte Carlo algorithms used to solve filtering problems arising in signal processing and Bayesian statistical inference. The filtering problem consists of estimating the internal states in dynamical systems when partial observations are made, and random perturbations are present in the sensors as well as in the dynamical system. The objective is to compute the posterior distributions of the states of some Markov process, given some noisy and partial observations. The term "particle filters" was first coined in 1996 by Del Moral [26] about mean field interacting particle methods used in fluid mechanics since the beginning of the 1960s. The terminology "sequential Monte Carlo" was proposed by Liu and Chen in 1998.

Particle filters' key advantage is their ability to represent arbitrary probability densities. Furthermore, unlike Kalman filters, particle filters can converge to the true posterior even in non-Gaussian, nonlinear dynamic systems. Compared to grid-based approaches, particle filters are very efficient because they automatically focus their resources (particles) on regions in state space with high probability. Because particle filters' efficiency strongly depends on the number of samples used for filtering, several improvements have been made to use the available samples more efficiently [27]. However, because these methods' worst-case complexity grows exponentially in the dimensions of the state space, we must be careful when applying particle filters to high-dimensional estimation problems.

1.7   Byzantine's Fault Tolerance and Brooks–Iyengar Hybrid Algorithm

1.7.1   Byzantine's Fault Tolerance

There are various algorithms and concepts developed and published to tolerate the imperfect information due to faulty or inaccurate nodes. One of the most famous among them for a distributed computing system is the Byzantine fault tolerance (BFT). BFT can be termed as the dependability of a fault tolerant distributed system where there is a possibility of failure of components leading to imperfect information being transmitted to different observers. This makes the decision of determining whether the information is valid or invalid difficult to judge and hence not come to a common conclusion on using the data or not. BFT comes a derivative of the Byzantine generals problem wherein the generals (actors) must come to an agreement on a commonly devised strategy in order to avoid any kind of catastrophic failure of the system. This has to be accomplished even when some of the generals are compromised and are no longer reliable. There are various other ways of addressing the BFT some of which are interactive inconsistency or source congruency, error avalanche, etc. More about the Byzantine generals problem and the Byzantine fault tolerance are discussed in Chaps. 2 and 3 of this book.

One of the applications of BFT in modern days is its use in crypto-currencies. Bitcoins which is a peer-to-peer digital currency system has a blockchain system to generate proof-of-work chains which form the key to overcome the Byzantine failures. Aircraft manufacturing giants like Boeing use the BFT approach in their control systems to tolerate failures and faults. Even spacecraft like the SpaceX Dragon flight system has employed BFT in its flight system design. The BFT technique uses the redundant messages or data in masking the failures and improving the systems' fault tolerance.

1.7.2 Brooks–Iyengar Hybrid Algorithm

This book has been written to highlight the Brooks–Iyengar hybrid algorithm and its various real-life applications. The Brooks–Iyengar hybrid algorithm is a seminal work that brought about a change in the way things were perceived in 1996 with distributed sensor networks. Even to date, this algorithm finds many applications some of which are discussed in the due course of this book. The Brooks–Iyengar hybrid algorithm for distributed systems, that are working in the presence of noisy or inaccurate data, combined the Byzantine agreement discussed above with sensor fusion and seamlessly bridges the gap between them. The algorithm can be touted to be the ideal mix of the famous Dolev's algorithm with the Mahaney and Schneider's fast convergence algorithm (FCA). The algorithm is efficient and runs in $O(N \log N)$ time. Some of the many applications of this algorithm include distributed control, high performance computing, software reliability, etc.

The Brooks–Iyengar algorithm is a run in every processing element (PE) of a distributed sensor network. Every PE exchanges its measured interval value with all other PEs in the network. The received measurement data from all the PEs are fused to find the weighted average of the midpoints of the region [28]. The working of the Brooks–Iyengar algorithm is depicted below. The algorithm is run on each PE individually:

Input: The value sent by PE k to PE i is a closed interval $[l_{k,i}, h_{k,i}]$, $1 \le k \le N$.
Output: The output of PE i includes a point estimate and an interval estimate:

1. PE i receives measurements from all the other PEs.
2. Divide the union of collected measurements into mutually exclusive intervals based on the number of measurements that intersect, which is known as the weight of the interval.
3. Remove intervals with weight less than $N - \tau$ where τ is the number of faulty PEs.

4. If there are L intervals left, let A_i denote the set of the remaining intervals. We have $A_i = \{(I_1^i, w_1^i), \dots, (I_L^i, w_L^i)\}$, where interval $I_L^i = [l_{I_l^i}, h_{I_l^i}]$ and w_l^i is the weight associated with interval I_l^i. We also assume $h_{I_l^i} \leq h_{I_{l+1}^i}$.

5. Calculate the point estimate v_i' of PE i as

$$v_i' = \frac{\sum_l \frac{l_{I_l^i} + h_{I_l^i}}{2}}{\sum_l w_l^i}$$

and the interval estimate is $[l_{I_l^i}, h_{I_l^i}]$.

Some of the characteristics of the algorithm are [28]:

- Faulty PEs tolerated $< N/3$
- Maximum faulty PEs $< 2N/3$
- Complexity $= O(N \log N)$
- Order of network bandwidth $= O(N)$
- Convergence $= 2t/N$
- Accuracy = limited by input
- Iterates for precision = often
- Precision over accuracy = no
- Accuracy over precision = no

A more detailed explanation and proof of the above algorithm is provided the upcoming sections of this book.

The Brooks–Iyengar algorithm is highly flexible and could be implemented in any network environment. The major milestone achieved by this algorithm was its use in MINIX to improve accuracy and precision thus leading to the flagship version of RT-Linux in 1996. The DARPA's SensIT program also used this algorithm for various distributed tracking systems. More about these applications are discussed in Sect. 1.2. Along with the above-mentioned applications, the Brooks–Iyengar algorithm also finds applications in the fields of robotic convergence, data and sensor fusion, safety of cyber-physical systems and ubiquitous computing, time-triggered architecture, ensemble learning in artificial intelligence systems to name a few. The algorithm has also been a part of the curriculum and taught in many universities like Purdue University, University of Maryland, Clemson University, University of Wisconsin, etc. The algorithm has been cited and used in more than 28 dissertation works over the years.

Even to date, there are new applications that have been identified to benefit from the ideas discussed in the Brooks–Iyengar algorithm which improves both the precision and accuracy of the interval measurements retrieved by a distributed sensor networks and making them robust enough to tolerate faulty sensor data. In 2016, the precision and accuracy bound of this algorithm was proved.

Figure 1.7 showcases how sensor fusion one of the pillars of the highly proclaimed Brooks–Iyengar algorithm encompasses the various distributed sensor networks and thus finds its use in most of the applications that require fault tolerance.

Fig. 1.7 The various fields in
which Sensor Fusion is used

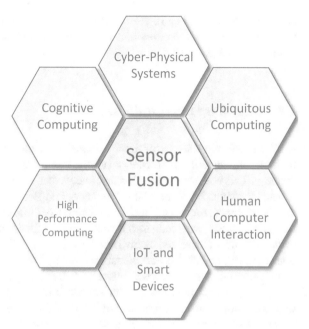

An area that is fast growing and is a good example of real-time processing of data from a fusion of distributed sensors is the modern automotive industry. The modern cars come equipped with a plethora of sensors each with a different function but with the final motive of helping in a hassle-free driving experience. The more recent trend of building and marketing autonomous self-driving cars would indeed use many sensors. Chapter 9 gives a brief review of the state of the art and about sensor fusion in autonomous cars. Autonomous cars are an interesting field since the precision required is very high and that too for a real-time processing the absence of which could have catastrophic impacts. Chapter 8 also provides an insight into the need for a robust fault tolerant safety-critical transportation application.

The book is organized in four parts that highlight the advances in distributed sensor fusion algorithms specifically focusing on the trends, the advances, and the future prospects. Part I gives a brief introduction to sensor networks, distributed networks, sensor fusion, and the need for fault tolerance. Some of the algorithms have also been theoretically analyzed. Part II focusses on the Advances of Sensor Fusion Algorithm wherein a theoretical analysis of the above described Brooks–Iyengar algorithm and an introduction to wireless sensor networks is provided. The various algorithms pertaining to the wireless sensor networks are also highlighted and analyzed along with the profound contributions of the Brooks–Iyengar algorithm in this cause. Part III highlights the trends of the Brooks–Iyengar algorithm by displaying one of its earliest applications dating back to 1996 wherein it was used in RT-MINIX and coming down to describing various other applications in the current day. Part IV puts forth the future prospects for the Brooks–Iyengar algorithm in the coming 10 years and the possibility of fusing the techniques with the modern-day technologies like blockchains, machine learning, deep learning, artificial intelligence, etc.

1.8 Summary and Outlook

A distributed wireless sensor network consists of thousands of sensor nodes, each with limited capabilities in energy supply, computer power, communications, and memory. This chapter introduces sensor networks, their numerous applications, distributed systems, Byzantine fault tolerance, and Brooks–Iyengar hybrid algorithm that is designed for sensor fusion. A formal definition for sensor fusion and its advantages were provided. In the first section of this chapter, we defined sensor networks and categorized them based on their applications into environmental branch, health branch, home automation branch, and other commercial applications. In the remaining sections, we introduced the need for fault tolerance in sensor networks and its important role in sustainability of sensor networks. Then, we addressed how sensor fusion and different filter types can help sensory data to become more precise in a distributed sensor network. Finally, we introduced the Byzantine fault tolerance method and the Brooks–Iyengar hybrid algorithm that would be discussed in further detail along the course of this book.

References

1. I.F., Akyildiz, W. Su, Y. Sankarasubramaniam, E. Cayirci, Wireless sensor networks: a survey. Comput. Netw. **38**(4), 393–422 (2002)
2. M.H. Amini, Distributed computational methods for control and optimization of power distribution networks, PhD Dissertation, Carnegie Mellon University, 2019
3. M.H. Amini, J. Mohammadi, S. Kar, Distributed holistic framework for smart city infrastructures: tale of interdependent electrified transportation network and power grid. IEEE Access **7**, 157535–157554 (2019)
4. A. Imteaj, M.H. Amini, Distributed sensing using smart end-user devices: pathway to federated learning for autonomous IoT, in *Proceeding of 2019 International Conference on Computational Science and Computational Intelligence*, Las Vegas (2019)
5. A. Imteaj, M.H. Amini, J. Mohammadi. Leveraging decentralized artificial intelligence to enhance resilience of energy networks (2019). arXiv preprint:1911.07690
6. E. Shih, S.-H. Cho, N. Ickes, R. Min, A. Sinha, A. Wang, A. Chandrakasan, Physical layer driven protocol and algorithm design for energy-efficient wireless sensor networks, in *Proceedings of the 7th Annual International Conference on Mobile Computing and Networking* (ACM, New York, 2001), pp. 272–287
7. J. Agre, L. Clare, An integrated architecture for cooperative sensing networks. Computer **33**(5), 106–108 (2000)
8. N. Bulusu, D. Estrin, L. Girod, J. Heidemann, Scalable coordination for wireless sensor networks: self-configuring localization systems, in *International Symposium on Communication Theory and Applications (ISCTA 2001), Ambleside, UK* (2001)

9. S.H. Cho, A.P. Chandrakasan, Energy efficient protocols for low duty cycle wireless microsensor networks, in *2001 IEEE International Conference on Acoustics, Speech, and Signal Processing, Proceedings (ICASSP'01)*, vol. 4 (IEEE, Piscataway, 2001), pp. 2041–2044

10. D. Estrin, Embedding the Internet. University of Southern Calif, 1999

11. I.A. Essa, Ubiquitous sensing for smart and aware environments. IEEE Pers. Commun. **7**(5), 47–49 (2000)

12. P. Johnson, D.C. Andrews, Remote continuous physiological monitoring in the home. J. Telemed. Telecare **2**(2), 107–113 (1996)

13. G.J. Pottie, W.J. Kaiser, Wireless integrated network sensors. Commun. ACM **43**(5), 51–58 (2000)

14. E.M. Petriu, N.D. Georganas, D.C. Petriu, D. Makrakis, V.Z. Groza, Sensor-based information appliances. IEEE Instrum. Meas. Mag. **3**(4), 31–35 (2000)

15. J. Rabaey, J. Ammer, J.L. Da Silva, D. Patel, Picoradio: Ad-hoc wireless networking of ubiquitous low-energy sensor/monitor nodes, in *IEEE Computer Society Workshop on VLSI, 2000, Proceedings* (IEEE, Piscataway, 2000), pp. 9–12

16. N.B. Priyantha, A. Chakraborty, H. Balakrishnan, The cricket location-support system, in *Proceedings of the 6th Annual International Conference on Mobile Computing and Networking* (ACM, New York, 2000), pp. 32–43

17. C.-C. Shen, C. Srisathapornphat, C. Jaikaeo, Sensor information networking architecture and applications. IEEE Pers. Commun. **8**(4), 52–59 (2001)

18. B. Walker, W. Steffen, An overview of the implications of global change for natural and managed terrestrial ecosystems. Conserv. Ecol. **1**(2), 1–14 (1997)

19. J.M. Kahn, R.H. Katz, K.S.J. Pister, Next century challenges: mobile networking for smart dust, in *Proceedings of the 5th Annual ACM/IEEE International Conference on Mobile Computing and Networking* (ACM, New York, 1999), pp. 271–278

20. M.H. Amini et al, Load management using multi-agent systems in smart distribution network, in *IEEE Power and Energy Society General Meeting* (IEEE, Piscataway, 2013), pp. 1–5

21. M.H. Amini (ed.), *Optimization, Learning, and Control for Interdependent Complex Networks*. Advances in Intelligent Systems and Computing, vol. 2 (Springer, Cham, 2020)

22. G.D. Abowd, J.P.G. Sterbenz, Final report on the inter-agency workshop on research issues for smart environments. IEEE Pers. Commun. **7**(5), 36–40 (2000)

23. W. Elmenreich, Sensor fusion in time-triggered systems, PhD Dissertation, Technischen Universit at Wien, 2002

24. V. Fox, J. Hightower, L. Liao, D. Schulz, G. Borriello, Bayesian filtering for location estimation. IEEE Pervasive Comput. **2**(3), 24–33 (2003)

25. D. Fox, W. Burgard, S. Thrun, Markov localization for mobile robots in dynamic environments. J. Artif. Intell. Res. **11**, 391–427 (1999)

26. P. Del Moral, Non-linear filtering: interacting particle resolution. Markov Process. Related Fields **2**(4), 555–581 (1996)

27. D. Estrin, R. Govindan, J. Heidemann, S. Kumar, Next century challenges: scalable coordination in sensor networks, in *Proceedings of the 5th Annual ACM/IEEE International Conference on Mobile Computing and Networking* (ACM, New York, 1999), pp. 263–270

28. S. Sahni, X. Xu, Algorithms for wireless sensor networks. University of Florida, Gainesville (September 7, 2004). Retrieved 23 Mar 2010

Chapter 2
Introduction to Algorithms for Wireless Sensor Networks

Overview: In the first section, the algorithms regarding sensor development and coverage are discussed. Among them, Kar and Banerjee's algorithm is explained in detail. Also, we discuss about how to evaluate the quality of sensor deployment. In the second section, we discuss the sensor networks routing algorithms based on unicast, multicast, and broadcast. Moreover, the algorithms regarding data collection and distribution are studied. Finally, third section focuses on sensor fusion algorithms including Brooks–Iyengar algorithm.

2.1 Introduction

A wireless sensor network may comprise thousands of sensor nodes. Each sensor node has a sensing capability as well as limited energy supply, compute power, memory, and communication ability. Besides military applications, wireless sensor networks may be used to monitor microclimates and wildlife habitats [1], the structural integrity of bridges and buildings, building security, location of valuable assets (via sensors placed on these valuable assets), traffic, and so on. However, realizing the full potential of wireless sensor networks poses myriad research challenges ranging from hardware and architectural issues, to programming languages and operating systems for sensor networks, to applications in smart cities infrastructures such as distributed processing [2–4] and network resilience [5], to security concerns, to algorithms for sensor network deployment, operation, and management. Iyengar

---

The following article with permission has been reproduced from the original copy: D. Tian, N.D. Georganas, A coverage-preserving node scheduling scheme for large wireless sensor networks, in *Proceedings of the 1st ACM International Workshop on Wireless Sensor Networks and Applications* (ACM, New York, 2002), pp. 32–41.

© Springer Nature Switzerland AG 2020   19
P. Sniatala et al., *Fundamentals of Brooks–Iyengar Distributed Sensing Algorithm*,
https://doi.org/10.1007/978-3-030-33132-0_2

and Brooks [6, 7] and Culler and Hong [8] provide good overviews of the breadth of sensor network research topics as well as of applications for sensor networks.

This chapter focuses on some of the algorithmic issues that arise in the context of wireless sensor networks. Specifically, we review algorithmic issues in sensor deployment and coverage, routing, and fusion. There is an abundance of algorithmic research related to wireless sensor networks. At a high level, the developed algorithms may be categorized as either centralized or distributed. Because of the limited memory, compute and communication capability of sensors, distributed algorithms research has focused on localized distributed algorithms—distributed algorithms that require only local (e.g., nearest neighbor) information.

2.2 Sensor Deployment and Coverage

In a typical sensor network application, sensors are to be placed (or deployed) so as to monitor a region or a set of points. In some applications we may be able to select the sites where sensors are placed while in others (e.g., in hostile environments) we may simply scatter (e.g., air drop) a sufficiently large number of sensors over the monitoring region with the expectation that the sensors that survive the air drop will be able to adequately monitor the target region. When site selection is possible, we use deterministic sensor deployment and when site selection is not possible, the deployment is nondeterministic. In both cases, it often is desirable that the deployed collection of sensors be able to communicate with one another, either directly or indirectly via multihop communication. So, in addition to covering the region or set of points to be sensed, we often require the deployed collection of sensors to form a connected network. For a given placement of sensors, it is easy to check whether the collection covers the target region or point set and also whether the collection is connected. For the coverage property, we need to know the sensing range of individual sensors (we assume that a sensor can sense events that occur within a distance r, where r is the sensor's sensing range, from it) and for the connected property, we need to know the communication range, c, of a sensor. Zhang and Lou [9] have established the following necessary and sufficient condition for coverage to imply connectivity.

Theorem 2.1 (Zhang and Lou [9]) *When the sensor density (i.e., number of sensors per unit area) is finite, $c \geq 2r$ is a necessary and sufficient condition for coverage to imply connectivity.*

Wang et al. [10] prove a similar result for the case of k-coverage (each point is covered by at least k sensors) and k-connectivity (the communication graph for the deployed sensors is k connected).

Theorem 2.2 (Wang et al. [10]) *When $c \geq 2r$, k-coverage of a convex region implies k-connectivity.*

Notice that k-coverage with $k > 1$ affords some degree of fault tolerance, we are able to monitor all points so long as no more than $k - 1$ sensors fail. Huang and Tseng [11] develop algorithms to verify whether a sensor deployment provides k-coverage. Other variations of the sensor deployment problem also are possible. For example, we may have no need for sensors to communicate with one another. Instead, each sensor communicates directly with a base station that is situated within the communication range of all sensors. In another variant [12, 13], the sensors are mobile and self-deployable. A collection of mobile sensors may be placed into an unknown and potentially hazardous environment. Following this initial placement, the sensors relocate so as to obtain maximum coverage of the unknown environment. They communicate the information they gather to a base station outside of the environment being sensed. A distributed potential-field-based algorithm to self-deploy mobile sensors under the stated assumptions is developed in [13] and a greedy and incremental self-deployment algorithm is developed in [12]. A virtual-force algorithm to redeploy sensors so as to maximize coverage also is developed by Zou and Chakrabarty [14]. Poduri and Sukhatme [15] develop a distributed self-deployment algorithm that is based on artificial potential fields and which maximizes coverage while ensuring that each sensor has at least k other sensors within its communication range.

2.2.1 Deterministic Deployment

2.2.1.1 Region Coverage

Kar and Banerjee [16] examine the problem of deploying the fewest number of homogeneous sensors so as to cover the plane with a connected sensor network. They assume that the sensing range equals the communication range (i.e., $r = c$). Algorithm 2.1 gives their deployment algorithm.

Algorithm 2.1 Kar and Banerjee's sensor deployment algorithm

Step 1 (achieve coverage)   Let $\delta = \left(\frac{\sqrt{3}}{2} + 1\right)r$. Place a sensor at $(i, j\delta)$, i even and j integer as well as one at $(i + r/2, j\delta)$, i odd and j integer.

Step 2 (achieve connectivity)   Let $\beta = \frac{\sqrt{3}}{2}r$. Place a sensor at $(0, j\delta \pm \beta)$, j odd.

One may verify that the sensors deployed in Step 1 are able to sense the entire plane. So, these sensors satisfy the coverage requirement. However, the sensors placed in Step 1 define many rows of connected sensors with the property that two sensors in different rows are unable to communicate (i.e., there is no multihop path between the sensors such that adjacent sensors on this path are at most c apart). Step 2 creates a connected network by placing a column of sensors in such a way as to connect together the connected rows that result from Step 1.

Kar and Banerjee [16] have shown that Algorithm 2.1 has a sensor density that is within 2.6% of the optimal density. This algorithm may be extended to provide connected coverage for a set of finite regions [16].

Algorithm 2.2 Greedy algorithm of [16] to deploy sensors

Step 1 (initialization)   Let s be any leaf of the Euclidean minimum-cost spanning tree of the point set. candidateSet $= \{s\}$

Step 2 (deploy sensor)   while (candidateSet $\neq \emptyset$) {

Remove any point p from candidateSet. Place a sensor at p. Remove from candidateSet all points covered by the sensor at p. Add to candidateSet all points (not necessarily vertices) q on the spanning tree T that satisfy the conditions:

1. q is distance r from p.
2. q is not covered by an already placed sensor.
3. The spanning tree path from s to q is completely covered by already placed sensors.

}

2.2.1.2   Point Coverage

Algorithm 2.2 gives the greedy algorithm of Kar and Banerjee [16] to deploy a connected sensor network so as to cover a set of points in Euclidean space. This algorithm, which assumes that $r = c$, uses at most 7.256 times the minimum number of sensors needed to cover the given point set [16]. It is easy to see that the constructed deployment covers all of the given points and is a connected network.

Grid coverage is another version of the point coverage problem. In this version, Chakrabarty et al. [17], we are given a two- or three-dimensional grid of points that are to be sensed. Sensor locations are restricted to these grid points and each grid point is to be covered by at least m, $m \geq 1$, sensors (i.e., we seek m-coverage). For sensing, we have k sensor types available. A sensor of type i costs c_i dollars and has a sensing range r_i. At most one sensor may be placed at a grid point. In this version of the point coverage problem, the sensors do not communicate with one another and are assumed to have a communication range large enough to reach the base station from any grid position. So, network connectivity is not an issue. The objective is to find a least-cost sensor deployment that provides m-coverage.

Chakrabarty et al. [17] formulate this m-coverage deployment problem as an integer linear program (ILP) with $O(kn^2)$ variables and $O(kn^2)$ equations, where n is the number of grid points. Xu and Sahni reduce the number of variables to $O(kn)$ and the number of equations to $O(n)$. Also, their formulation does not require the sensor locations and points to be sensed to form a grid. Let s_{ij} be a 0/1-valued variable with the interpretation s_{ij} iff a sensor of type i is placed at point j, $1 \leq i \leq k$, $1 \leq j \leq n$. The solution to the following ILP describes an optimal sensor deployment.

$$\text{minimize} \sum_{i=1}^{k} \sum_{j=1}^{n} c_i s_{ij}$$

$$\forall j = 1, 2, \ldots, n, \quad \sum_{i=1}^{k} \sum_{a \in X(i,j)} s_{ia} \geq m$$

$$\forall j = 1, 2, \ldots, n, \quad \sum_{i=1}^{k} s_{ij} \leq 1$$

where $X(i, j)$ is the set of all points within r_i of point j.

Even with this reduction in the number of variables and equations, the ILP is practically solvable only for a small number of points n. For large n, Chakrabarty et al. [17] propose a divide-and-conquer "near-optimal" algorithm in which the base case (small number of points) is solved optimally using the ILP formulation.

2.2.2 Maximizing Coverage Lifetime

When sensors are deployed in difficult-to-access environments, as is the case in many military applications, a large number of sensors may be air-dropped into the region that is to be sensed. Assume that the sensors that survive the air drop cover all targets that are to be sensed. Since the power supply of a sensor cannot be replenished, a sensor becomes inoperable once it runs out of energy. Define the life of a sensor network to be the earliest time at which the network ceases to cover all targets. The life of a network can be increased if it is possible to put redundant sensors (i.e., sensors not needed to provide coverage of all targets) to sleep and awaken these sleeping sensors when they are needed to restore target coverage. Sleeping sensors are inactive while sensors that are awake are active. Inactive sensors consume far less energy than active ones.

Cardei and Du [18] propose partitioning the set of available sensors into disjoint sets such that each set covers all targets. Let T be the set of targets to be monitored and let S_i denote the subset of T in the range of sensor i, $1 \leq i \leq n$. Let $P_1, P_2, \ldots, P_k$ be disjoint partitions of the set of n sensors such that $\bigcup_{j \in P_i} S_j = T$ $1 \leq i \leq k$. Then the set of sensors in each P_i covers all targets. We refer to the set of P_is as a disjoint set cover of size k. Moreover, by going through k sleep/awake rounds where in round i only the sensors in P_i are awake, we are able to monitor all targets in each round and increase the network life to almost kt, where t is the time it takes a sensor to deplete its energy when in the awake mode. Since sensors deplete energy even when in the sleep mode, the life of each round is slightly less than that of the preceding round. Cardei and Du [18] have shown that deciding whether there is a disjoint set cover of size k for a given sensor set is NP-complete. They develop a heuristic to maximize the size of a disjoint set cover. An experimental evaluation of this heuristic reveals that it finds about 10% more disjoint covers than does the best algorithm of Slijepcevic and Potkonjak [19]. However, the algorithm of Cardei and Du [18] takes more time to execute.

Several decentralized localized protocols to control the sleep/awake state of sensors so as to increase network lifetime have been proposed. Ye et al. [20] propose a very simple protocol. In this protocol, the set of active nodes provide the desired coverage. A sleeping node wakes up when its sleep timer expires and broadcasts a probing signal a distance d (d is called the probing range). If no active sensor is detected in this probing range, the sensor moves into the active state. However, if an active sensor is detected in the probing range, the sensor determines how long to sleep, sets its sleep timer, and goes to sleep. Techniques to dynamically control the sleep time and probing range are discussed in [20]. Simulations reported in [20] indicate that this simple protocol outperforms the GAF protocol of [21]. However, experiments conducted by Tian and Georganas [22] reveal that the protocol of Ye et al. [20] "cannot ensure the original sensing coverage and blind spots may appear after turning off some nodes." Tian and Georganas [22] propose an alternative distributed localized protocol that sensors may use to turn themselves on and off. The network operates in rounds, where each round has two phases—self-scheduling and sensing. In the self-scheduling phase each sensor decides whether or not to go to sleep. In the sensing phase, the active/awake sensors monitor the region. Sensor s turns itself off in the self-scheduling phase if its neighbors are able to monitor the entire sensing region of s. To make this determination, every sensor broadcasts its location and sensing range. A backoff scheme is proposed to avoid blind spots that would otherwise occur if two sensors turn off simultaneously, each expecting the other to monitor part or all of its sensing region. In this backoff scheme, each active sensor uses a random delay before deciding whether or not it can go to sleep without affecting sensing coverage.

The decentralized algorithm OGDC (optimal geographical density control) of Zhang and Lou [9] guarantees coverage, which by Theorem 2.1 implies connectivity whenever the sensor communication range is at least twice its sensing range. Experimental results reported in [9] suggest that when OGDC is used, the number of active (awake) nodes may be up to half that when the PEAS [23] or GAF [21] algorithms are used to control sensor state. The coverage configuration protocol (CCP) to maximize lifetime while providing k-coverage (as well as k-connectivity when $c \geq 2r$, Theorem 2.2) is developed in [10]. A distributed protocol for differentiated surveillance is proposed by Yan et al. [24]. Assume that the probability that sensor s detects an event at a distance d is $P(s, x) = 1/(1 + \alpha d)^\beta$. The probability $P(x)$ that an event at x is detected by the sensor network is

$$P(x) = 1 - \Pi(1 - P(s, x))$$

where the product is taken over all sensors in the network. The coverage, C, of the sensor network (overall network coverage) is the sum of $P(x)$ over all points in the sensing region. Let $C'(s)$ be the overall coverage when sensor s is removed from the network. The sensing denomination [25] of a sensor s is its contribution, $C - C'(s)$, to overall network coverage. Lu and Suda [25] use the sensing denomination of a sensor to obtain a distributed localized self-scheduling algorithm to schedule the sleep/awake states of sensors so as to increase network lifetime. In their algorithm, each sensor periodically makes a decision to either go to sleep or be active. Sensors with higher sensing denomination have a higher probability of being active.

2.2.3   Deployment Quality

The quality of a sensor deployment may be measured by the minimum k for which the deployment provides k-coverage as well as by the minimum k for which we have k-connectivity. By Theorem 2.2 the first metric implies the second when the communication range c is at least twice the sensing range r. Meguerdichian et al. [26] have formulated additional metrics suitable for a variety of sensor applications. In these applications, the ability of a sensor to detect an activity at distance d from the sensor is given by the function $1/(1 + \alpha d)^\beta$, where α and β are constants. So, sensing ability is maximum when $d = 0$ and declines as we get farther from the sensor.

Let P be a path that connects two points u and v (these points may be within or outside the region being sensed). The breach weight, $BW(P, u, v)$, of P is the closest path P gets to any of the deployed sensors. The breach weight, $BW(u, v)$, of the points u and v is the maximum of the breach weights of all paths between u and v.

$$BW(u, v) = \max\{BW(P, u, v) | P \text{ is a path between } u \& v\}$$

The breach weight or breachability, BW, of a sensor network is the maximum of the breach weights of all point pairs.

$$BW(u, v) = \max\{BW(u, v) | u \& v \text{ are points on the boundary of the sensing region}\}$$

When the ability of a sensor to detect an activity is inversely proportional to some power k of distance, sensor deployments that minimize breach weight are preferred. The breachability of a network gives us an indication of how successful an intruder could be in evading detection. Suppose we have an application in which items are being transported between pairs of points and our sensors track the progress of these shipments. Now we wish to use maximally observable paths, that is, paths that remain as close to a sensor as possible. Let $d(x)$ be the distance between a point x and the sensor nearest to x. Meguerdichian et al. [26] define the support weight, $SW(P, u, v)$, of a path P between u and v as

$$SW(P, u, v) = \max\{d(x) | x \text{ is a point on } P\}$$

$SW(u, v)$ now is defined as

$$SW(u, v) = \min\{SW(P, u, v) | P \text{ is a path between } u \& v\}$$

and the support weight (or simply support), SW, of the sensor network is

$$SW = \max\{SW(u, v) | u \& v \text{ are points on the boundary of the sensing region}\}$$

Although Meguerdichian et al. [26] develop centralized algorithms to compute $BW(u, v)$ and $SW(u, v)$, these algorithms are flawed [27]. Li et al. [27] describe a distributed localized algorithm to determine $SW(u, v)$. This algorithm is given in Algorithm 2.3. In this algorithm $|sa|$ is the Euclidean distance between the sensors s and a. The algorithm assumes that u and v are in the convex hull of the sensor locations.

Since there may be several paths with the computed $SW(u, v)$ value, we may be interested in finding, say, a best support path from u to v that has minimal length. Li et al. [27] develop a localized distributed algorithm to find an approximately minimal length path with support $SW(u, v)$.

The exposure $E(P, u, v)$ of a path P from u to v is defined as

$$E(P, u, v) = \int_u^v S(x) \mathrm{d}x$$

where $S(x)$ is a sensing function and the integral is computed over the path P. Meguerdichian et al. [28] suggest two sensing functions. One is simply the sensing ability of the closest sensor to x; the other is the sum of the abilities of all sensors to detect activity at x. An intruder who wishes to minimize the risk of detection would take a minimal-exposure path to get from u to v. Algorithm 2.4 proposed by [28] finds an approximation to the minimal-exposure path between u and v. This algorithm overlays an undirected weighted graph over the sensing region and then finds a shortest path from the graph vertex nearest to u to the graph vertex nearest to v. This shortest path may be found using Dijkstra's shortest path algorithm. By increasing the number of vertices and edges used in Step 1, the accuracy of the approximation is increased. For the overlay graph, [28] considers a uniform two-dimensional grid arrangement of the vertices and a variety of options for edges (e.g., edges connect a vertex to its north, south, east, west, northwest, northeast, southeast, and southwest neighbors). Sensor deployments that maximize minimal-exposure are preferred.

Algorithm 2.3 Distributed algorithm of [27] to compute $SW(u, v)$

Step 1 (construct local neighborhood graph)   Each sensor broadcasts its id and
    location. Each sensor s compiles a list $L(s)$ of all ids and locations that it hears.
    Let $A(s)$, the adjacency list for s, comprise all sensors $a \in L(s)$ such that there
    is no $b \in L(s)$ located in the interior of the intersection region of the radius $|sa|$
    circles centered at s and a. For each $a \in A(s)$, the weight of the edge (s, a) is
    $|sa|/2$.

Step 2 (construct best support path)   Let the length of a path be the maximum weight of its edges. Let x and y, respectively, be the sensors closest to the points u to v. Run the distributed Bellman–Ford shortest path algorithm to determine a shortest path $P(x, y)$, in the local neighborhood graph, from x to y. (u, x), $P(x, y)$, (y, v) is a best support path from u to v. The weight of (u, x) is $|ux|$ and that of (y, v) is $|yv|$. $SW(u, v)$ is the maximum of the edges weights in the best support path.

Algorithm 2.4   Algorithm of [28] to find an approximate minimal-exposure path

---

Step 1: (graph overlay)   Overlay the sensing region with a weighted undirected graph G. The vertices of G are points in the sensing region and its edges are straight lines. The weight of an edge is the exposure of that edge.

Step 2: (minimal-exposure path)   The minimal-exposure path from u to v is estimated to be the shortest path in G from the vertex of G closest to the point u to the vertex of G closest to the point v. Its exposure is the length of this shortest path.

Veltri et al. [29] develop a distributed localized algorithm to find an approximate minimal-exposure path. Also, they show that finding a maximal-exposure path is NP-hard and they propose several heuristics to construct approximate maximal-exposure paths. Additionally, a linear programming formulation for minimal- and maximal-exposure paths is obtained. Kanan et al. [30] develop polynomial time algorithms to compute the maximum vulnerability of a sensor deployment to attack by an intelligent adversary and use this to compute optimal deployments with minimal vulnerability.

2.3 Routing

Traditional routing algorithms for sensor networks are datacentric in nature. Given the unattended and untethered nature of sensor networks, datacentric routing must be collaborative as well as energy-conserving for individual sensors. Kannan et al. [31, 32] have developed a novel sensor-centric paradigm for network routing using game-theory. In this sensor-centric paradigm, the sensors collaborate to achieve common network-wide goals such as route reliability and path length while minimizing individual costs. The sensor-centric model can be used to define the quality of routing paths in the network (also called path weakness). Kannan et al. [32] describe inapproximability results on obtaining paths with bounded weakness along with some heuristics for obtaining strong paths. The development of efficient distributed algorithms for approximately optimal strong routing is an open issue that can be explored further.

Energy conservation is an overriding concern in the development of any routing algorithm for wireless sensor networks. This is because such networks are often located such that it is difficult, if not impossible, to replenish the energy supply of a sensor. Three forms—unicast, broadcast, and multicast—of the routing problem have received significant attention in the literature. The overall objective of these algorithms is to either maximize the lifetime (earliest time at which a communication fails) or the capacity of the network (amount of data traffic carried by the network over some fixed period of time).

Assume that the wireless network is represented as a weighted directed graph G that has n vertices/nodes and e edges. Each node of G represents a node of the wireless network. The weight $w(i, j)$ of the directed edge (i, j) is the amount of energy needed by node i to transmit a unit message to node j.

In the most common model used for power attenuation, signal power attenuates at the rate a/rd, where a is a media dependent constant, r is the distance from the signal source, and d is another constant between 2 and 4 [33]. So, for this model, $w(i, j) = w(j, i) = c \times r(i, j)d$, where $r(i, j)$ is the Euclidean distance between nodes i and j and c is a constant. In practice, however, this nice relationship between $w(i, j)$ and $r(i, j)$ may not apply. This may, for example, be due to obstructions between the nodes that may cause the attenuation to be larger than predicted. Also, the transmission properties of the media may be asymmetric resulting in $w(i, j) \neq w(j, i)$.

2.3.1 Unicast

In a unicast, we wish to send a message from a source sensor s to a destination sensor t. Singh et al. [34] propose five strategies that may be used in the selection of the routing path for this transmission. The first of these is to use a minimum energy path (i.e., a path in G for which the sum of the edge weights is minimum) from s to t. Such a path may be computed using Dijkstra's shortest path algorithm [35]. However, since, in practice, messages between several pairs of source–destination sensors need to be routed in succession, using a minimum energy path for a message may prevent the successful routing of future messages. As an example, consider the graph of Fig. 2.1. Suppose that sensors $x, b_1, \ldots, b_n$ initially have ten units of energy each and that $u_1, \ldots, u_n$ each have one unit. Assume that the first unicast is a unit-length message from x to y. There are exactly two paths from x to y in the sensor network of Fig. 2.1. The upper path, which begins at x, goes through each of the u_is, and ends at y uses $n + 1$ energy units; the lower path uses $2(n + 1)$ energy units. Using the minimum energy path, depletes the energy in node u_i, $1 \leq i \leq n$. Following the unicast, sensors $u_1, \ldots, u_n$ are unable to forward any messages. So an ensuing request to unicast from u_i to u_j, $i < j$ will fail. On the other hand, had we used the lower path, which is not a minimum energy path, we would not deplete the energy in any sensor and all unit-length unicasts that could be done in the initial network also can be done in the network following the first x to y unicast.

The remaining four strategies proposed in [34] attempt to overcome the myopic nature of the minimum energy path strategy, which sacrifices network lifetime and

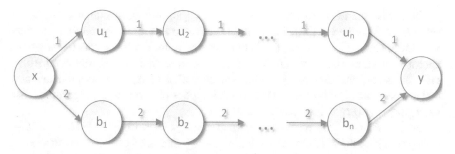

Fig. 2.1 A sensor network

capacity in favor of total remaining energy. Since routing decisions must be made in an online fashion (i.e., if the ith message is to be sent from s_i to t_i, the path for message i must be decided without knowledge of s_j and t_j, $j > i$), we seek an online algorithm with good competitive ratio. It is easy to see that there can be no online unicast algorithm with constant competitive ratio with respect to network lifetime and capacity [36]. For example, consider the network of Fig. 2.1. Assume that the energy in each node is one unit. Suppose that the first unicast is from x to y. Without knowledge of the remaining unicasts, we must select either the upper or lower path from x to y. If the upper path is chosen and the source–destination pairs for the remaining unicasts turn out to be $(u_1, u_2), (u_2, u_3), \ldots, (u_n, y)$ then the online algorithm routes only the first unicast whereas an optimal offline algorithm would route all $n + 1$ unicasts, giving a competitive ratio of $n + 1$. The same ratio results when the lower path is chosen and the source–destination pairs for the remaining unicasts are $(b_1, b_2), (b_2, b_3), \ldots, (b_n, y)$.

Theorem 2.3 *There is no online algorithm to maximize lifetime or capacity that has a competitive ratio smaller than $\Omega(n)$.*

To maximize lifetime and/or capacity, we need to achieve some balance between the energy consumed by a route and the minimum residual energy at the nodes along the chosen route. Aslam et al. [36] propose the max-min $z P_{\min}$-path algorithm to select unicast routes that attempt to make this balance. This algorithm selects a unicast path that uses at most $z \times P_{\min}$ energy, where z is a parameter to the algorithm and $P_{\min}$ is the energy required by the minimum energy unicast path. The selected unicast path maximizes the minimum residual-energy fraction (energy remaining after unicast/initial energy) for nodes on the unicast path. Notice that the possible values for the residual-energy fraction of node u may be obtained by computing $(ce(u) - w(u, v) \times l)/ie(u)$, where l is the message length, $ce(u)$ is the (current) energy at node u just before the unicast, $ie(u)$ is the initial energy at u, and $w(u, v)$ is the energy needed to send a unit-length message along the edge (u, v). This computation is done for all vertices v adjacent from u. Hence the union, L, of these values taken over all u gives the possible values for the minimum residual-energy fraction along any unicast path. Algorithm 2.5 represents the max-min $z P_{\min}$ algorithm.

Several adaptations to the basic max-min $z P_{\min}$ algorithm, including a distributed version are described in [36]. Kar et al. [37] develop an online capacity-competitive algorithm, CMAX, with logarithmic competitive ratio. On the surface, this would appear to violate Theorem 2.3. However, to achieve this logarithmic competitive ratio, the algorithm CMAX does admission control. That is, it rejects some unicasts that are possible. The bound of Theorem 2.3 applies only for online algorithms that must perform every unicast that is possible.

Let ce, E, and l be as for the max-min $z P_{\min}$ algorithm. Let $\alpha(u) = 1 - ce(u)/ie(u)$ be the fraction of u's initial energy that has been used so far. Let λ and σ be two constants. In the CMAX algorithm, the weight of every edge (u, v) is changed from $w(u, v)$ to $w(u, v) \times (\lambda^{\alpha(u)} - 1)$. The shortest source-to-destination path P in the resulting graph is determined. If the length of this path is more than σ, the unicast is rejected (admission control); otherwise, the unicast is done using path P. Algorithm 2.6 gives the details of this algorithm.

Algorithm 2.5 The max-min $z P_{\min}$ unicast algorithm of [38]

Step 1 (initialize)   Eliminate from G every edge (u, v) for which $ce(u) < w(u, v) \times l$.

Let L be the list of possible values for the minimum residual-energy fraction.

Step 2 (binary search)   Do a binary search in L to find the maximum value max of the minimum residual-energy fraction for which there is a path P from source to destination that uses at most $z \times P_{\min}$ energy.

For this, when testing a value q from L, we find a shortest source-to-destination path that does not use edges (u, v) that make the residual-energy fraction at u less than q.

Step 3 (wrap up)   If no path is found in Step 2, the unicast is not possible.

Otherwise, use the path P corresponding to max.

Algorithm 2.6 CMAX algorithm of [37] for unicasts

Step 1 (initialize)   Eliminate from G every edge (u, v) for which $ce(u) < w(u, v) \times l$.

Change the weight of every remaining edge (u, v) to $w(u, v) \times (\lambda^{\alpha(u)} - 1)$.

Step 2 (shortest path)   Let P be the shortest source-to-destination path in the modified graph.

Step 3 (wrap up)   If no path is found in Step 2, the unicast is not possible.

If the length of P is more than σ do not do the unicast.

Otherwise, use P for the unicast.

The CMAX algorithm of Algorithm 2.6 has a complexity advantage over the max-min $z P_{\min}$ algorithm of Algorithm 2.5. The former does only one shortest path computation per unicast while the latter does $O(\log n)$, where n is the number of sensor nodes. Although admission control is necessary to establish the logarithmic competitive ratio bound for CMAX, we may use CMAX without admission control (i.e., route very unicast that is feasible) by setting $\sigma = \infty$. Experimental results reported in [37] suggest that CMAX with no admission control outperforms max-min zP_{min} on both the lifetime and capacity metrics.

In the MRPC lifetime-maximization algorithm of Misra and Banerjee [39], the capacity, $c(u, v)$ of edge (u, v) is defined to be $ce(u)/w(u, v)$. Note that $c(u, v)$ is the number of unit-length messages that may be transmitted along (u, v) before node u runs out of energy. The lifetime of path P, life(P) is defined to be the minimum edge capacity on the path. In MRPC, the unicast is done along a path P with maximum lifetime. Algorithm 2.7 gives the MRPC unicast algorithm. A decentralized implementation as well as a conditional MRPC in which minimum energy routing is used so long as the selected path has a lifetime that is greater than or equal to a specified threshold. When the lifetime of the selected path falls below this threshold, we switch to MRPC routing.

Algorithm 2.7 MRPC algorithm of [39] for unicasts

Step 1 (initialize)   Eliminate from G every edge (u, v) for which $ce(u) < w(u, v) \times l$. For every remaining edge (u, v), let $c(u, v) = ce(u)/w(u, v)$. Let L be the list of distinct $c(u, v)$ values.

Step 2 (binary search)   Do a binary search in L to find the maximum value max for which there is a path P from source to destination that uses no edge with $c(u, v) < \text{max}$. For this, when testing a value q from L, we perform a depth- or breadth-first search beginning at the source. The search is not permitted to use edges with $c(u, v) < q$. Let P be the source-to-destination path with lifetime max.

Step 3 (wrap up)   If no path is found in Step 2, the unicast is not possible. Otherwise, use P for the unicast.

Chang and Tassiulas [40, 41] develop a linear programming formulation for lifetime maximization. This formulation requires knowledge of the rate at which each node generates unicast messages. Wu et al. [42] propose unicast routing based on connected dominating sets to maximize network lifetime. Stojmenovic and Lin [43] and Melodia et al. [44] develop localized unicast algorithms to maximize lifetime and Heinzelman et al. [45] develop a clustering-based routing algorithm (LEACH) for sensor networks.

2.3.2 Multicast and Broadcast

Using an omnidirectional antenna, node i can transmit the same unit message to nodes $j_1, j_2, \ldots, j_k$, using

$$e_{\text{wireless}} = \max\{w(i, j_q) | 1 \leq q \leq k\}$$

energy rather than

$$e_{\text{wired}} = \sum_{q=1}^{k} w(i, j_q)$$

energy. Since, $e_{\text{wireless}} \leq e_{\text{wired}}$, the reduction in energy needed to broadcast from one node to several others in a wireless network over that needed in a wired network is referred to as the wireless broadcast advantage [46, 47].

Because of the similarity between multicast and broadcast algorithms for wireless sensor networks, we need focus on just one of multicast and broadcast. In this chapter, our explicit focus is broadcast. To broadcast from a source s, we use a broadcast tree T, which is a spanning tree of G that is rooted at s. The energy, $e(u)$, required by a node of T to broadcast to its children is

$$e(u) = \max\{w(u, v)|v \text{ is a child of } u\}$$

Note that for a leaf node u, $e(u) = 0$. The energy, $e(T)$, required by the broadcast tree to broadcast a unit message from the source to all other nodes is

$$e(T) = \sum_{u} e(u)$$

For simplicity, we assume that only unit-length messages are broadcast. So, following a broadcast using the broadcast tree T, the residual energy, $re(i, T)$, at node i is

$$re(i, T) = ce(i) - \max\{w(i, j)|j \text{ is a child of } i \text{ in } T\} \geq 0$$

The problem, MEBT, of finding a minimum energy broadcast tree rooted at a node s is NP-hard, because MEBT is a generalization of the connected dominating set problem, which is known to be NP-hard [48]. In fact, MEBT cannot be approximated in polynomial time within a factor $(1 - \varepsilon)\Delta$, where ε is any small positive constant and Δ is the maximal degree of a vertex, unless $NP \subseteq DTIME[n^{O(\log \log n)}]$ [49]. Marks et al. [50] propose a genetic algorithm for MEBT and Das et al. [51] propose integer programming formulations.

Wieselthier et al. [46, 47] propose four centralized greedy heuristics—DSA, MST, BIP, BIPPN—to construct minimum energy broadcast trees. The DSA (Dijkstra's shortest paths algorithm) heuristic (this is the BLU of [46, 47] augmented with the sweep pass of [46, 47]) constructs a shortest path from the source node s to every other vertex in G. The constructed shortest paths are superimposed to obtain a tree T rooted at s. Finally, a sweep is performed over the nodes of T. In this sweep, nodes are examined in ascending order of their index (i.e., in the order $1, 2, \ldots, n$). The transmission energy $\tau(i)$ for node i is determined to be the

$$\max\{w(i, j)|j \text{ is a child of } i \text{ in } T\}$$

If using $\tau(i)$ energy, node i is able to reach any descendants other than its children, then these descendants are promoted in the broadcast tree T and become additional children of i.

The MST (minimum spanning tree) (this is the BLiMST heuristic of [46] augmented with a sweep pass) uses Prim's algorithm [35] to construct a minimum-cost spanning tree (the cost of a spanning tree is the sum of its edge weights). The constructed spanning tree is restructured by performing a sweep over the nodes to reduce the total energy required by the tree.

The BIP (broadcast incremental power) heuristic (this is the BIP heuristic of [46, 47] augmented with a sweep pass) begins with a tree T that comprises only the source node s. The remaining nodes are added to T one node at a time. The next node u to add to T is selected so that u is a neighbor of a node in T and $e(T \cup \{u\}) - e(T)$ is minimum. Once the broadcast tree is constructed, a sweep is done to restructure the tree so as to reduce the required energy.

The BIPPN (broadcast incremental power per node) heuristic (this is the node-based MST heuristic of [47] augmented with a sweep pass) begins with a tree T that comprises only the source node s and uses several rounds to grow T into a broadcast tree. To describe the growth procedure, we define $e(T, v, i)$, where $v \in T$, to be the minimum incremental energy (i.e., energy above the level at which v must broadcast to reach its present children in T) needed by node v so as to reach i of its neighbors that are not in T (of course, only neighbors j such that $ce(v) \geq w(v, j)$ are to be considered). Let $R(T, v, i) = i/e(T, v, i)$. Note that $R(T, v, i)$ is the inverse of the incremental energy needed per node added to T. In each round of BIPPN, we determine v and i such that $R(T, v, i)$ is maximum. Then, to T, we add the i neighbors of v that are not in T and can be reached from v by incrementing v's broadcast energy by $e(T, v, i)$. The i neighbors are added to T as children of v. Once the broadcast tree is constructed, a sweep is done to restructure the tree so as to reduce the required energy.

Wan et al. [52] show that when $w(i, j) = c \times rd$, the MST and BIP heuristics have constant approximation ratios and that the approximation ratio for DSA is at least $n/2$. Park and Sahni [53] describe two additional centralized greedy heuristics—MEN and BIPWLA—to construct minimum energy broadcast trees. The second of these (BIPWLA) is an adaptation of the modified greedy algorithm proposed by Guha and Khuller [49] for the connected dominating set problem.

The MEN (maximum energy node) heuristic attempts to use nodes that have more energy as non-leaf nodes of the broadcast tree (note that when i is a non-leaf, $re(i) < ce(i)$, whereas when i is a leaf, $re(i) = ce(i)$). In MEN, we start with $T = \{s\}$. At each step, we determine Q such that

$$Q = \{u | u \text{ is a leaf of } T \text{ and } u \text{ has a neighbor } j, j \notin T, \text{ for which } w(u, j) \leq ce(u)\}$$
$$(2.1)$$

From Q, we select the node u that has maximum energy $ce(u)$. All neighbors j of u not already in T and which satisfy $w(u, j) \leq ce(u)$ are added to T as children of u. This process of adding nodes to T terminates when T contains all nodes of G (i.e., when T is a broadcast tree). Finally, a sweep is done to restructure the tree so as to reduce the required energy.

The BIPWLA (broadcast incremental power with look ahead) heuristic is an adaptation of the look ahead heuristic proposed in [49] for the connected dominating set problem. This heuristic also may be viewed as an adaptation of BIPPN. In BIPWLA, we begin with a tree T that comprises the source node s together with all neighbors of s that are reachable from s using $ce(s)$ energy. Initially, the source node s is colored black, all other nodes in T are gray and nodes not in T are white. Nodes not in T are added to T in rounds. In a round, one gray node will have its color changed to black and one or more white nodes will be added to T as gray nodes. It will always be the case that a node is gray iff it is a leaf of T, it is black iff it is in T but not a leaf, it is white iff it is not in T. In each round, we select one of the gray nodes g in T; color g black; and add to T all white neighbors of g that can be reached using $ce(g)$ energy. The selection of g is done in the following manner. For each gray node $u \in T$, let n_u be the number of white neighbors of u reachable from u by a broadcast the uses $ce(u)$ energy. Let p_u be the minimum energy needed to reach these n_u nodes by a broadcast from u. Let,

$$A(u) = \{j | w(u, j) \le ce(u) \text{ and } j \text{ is a white node}\}$$

We see that $n_u = |A(u)|$ and $p_u = \max\{w(u, j) | j \in A(u)\}$.

For each $j \in A(u)$, we define the following analogous quantities $B(j) = \{q | w(j, q) \le ce(j) \text{ and } q \text{ is a white node}\}$, $n_j = |B(j)|$, and $pj = \max\{w(j, q) | q \in B(j)\}$. Node g is selected to be the gray node u of T that maximizes

$$n_u/p_u + \max\{n_j/p_j | j \in B(u)\}.$$

Once the broadcast tree is constructed, a sweep is done to restructure the tree so as to reduce the required energy.

A localized distributed heuristic for MEBT that is based on relative neighborhood graphs has been proposed by Cartigny et al. [54]. Wu and Dai [55] develop an alternative localized distributed heuristic that uses k-hop neighborhood information.

In a real application, the wireless network will be required to perform a sequence $B = b_1, b_2, \ldots$ of broadcasts. Broadcast b_i will specify a source node s_i and a message length l_i. Assume, for simplicity, that $l_i = 1$ for all i. For a given broadcast sequence B, the network lifetime is the largest i such that broadcasts $b_1, b_2, \ldots, b_i$ are successfully completed. The six heuristics of [46, 47, 53] may be used to maximize lifetime by performing each b_i using the broadcast tree generated by the heuristic (the broadcast trees are generated in sequence using the residual node energies). However, since each of these heuristics is designed to minimize total energy consumed by a single broadcast, it is entirely possible that the very first broadcast depletes the energy in a node, making subsequent broadcasts impossible.

Singh et al. [56] propose a greedy heuristic for a broadcast sequence. The source node broadcasts to each of its neighbors resulting in a 2-level tree T with the source as root. T is expanded into a broadcast tree through a series of steps. In each step, a leaf of T is selected and its non-tree network neighbors added to T. The leaf selection is done using a greedy strategy—select the leaf for which the ratio (energy expended by this leaf so far)/(number of non-tree leaves) is minimum.

The critical energy, $CE(T)$, following a broadcast that uses the broadcast tree T is defined to be

$$CE(T) = \min\{re(i, T)|1 \leq i \leq n\}$$

Park and Sahni [53] suggest the use of broadcast trees that maximize the critical energy following each broadcast. Let $MCE(G, s)$ be the maximum critical energy following a broadcast from node s. For each node i of G, define $a(i)$ as below:

$$a(i) = \{ce(i) - w(i, j)|(i, j) \text{ is an edge of } G \text{ and } ce(i) \geq w(i, j)\}$$

Let $l(i)$ denote the set of all possible values for $re(i)$ following the broadcast. We see that

$$l(i) = \begin{cases} a(i) & \text{if } i = s \\ a(i) \cup \{ce(i)\} & \text{otherwise} \end{cases}$$

Consequently, the sorted list of all possible values for $MCE(G, s)$ is given by

$$L = \text{sort} \left(\bigcup_{i=1}^{n} l(i) \right)$$

We may determine whether G has a broadcast tree rooted at s such that $CE(T) \geq q$ by performing either a breadth-first or depth-first search [35] starting at vertex s. This search is forbidden from using edges (i, j) for which $ce(i) - w(i, j) < q$. $MCE(G, s)$ may be determined by performing a binary search over the values in L [53].

Each of the heuristics described in [46, 47, 53] to construct a minimum energy broadcast tree may be modified so as to construct a minimum energy broadcast tree T for which $CE(T) = MCE(G, s)$. For this modification, we first compute $MCE(G, s)$ and then run a pruned version of the desired heuristic. In this pruned version, the use of edges for which $ce(i) - w(i, j) < MCE(G, s)$ is forbidden. Experiments reported in [53] indicate that this modification significantly improves network lifetime, regardless of which of the six base heuristics is used. Lifetime improved, on average, by a low of 48.3% for the MEN heuristic to a high of 328.9% for the BIPPN heuristic. The BIPPN heuristic modified to use $MCE(G, s)$, as above, results in the best lifetime.

2.3.3 Data Collection and Distribution

In the data collection problem, a base station is to collect sensed data from all deployed sensors. The data distribution problem is the inverse problem in which the base station has to send data to the deployed sensors (different sensors receive different data from the base station). In both of these problems, the objective is to complete the task in the smallest amount of time. Florens and McEliece [38, 57] have observed that the data collection and distribution problems are symmetric. Hence, once can derive an optimal data collection algorithm from an optimal data distribution algorithm and vice versa. Therefore, it is necessary to study just one of these two problems explicitly. In keeping with [38, 57], we focus on data distribution.

Let $S_1, \ldots, S_n$ be n sensors and let S_0 represent the base station. Let p_i be the number of data packets the base station has to send to sensor i, $1 \leq i \leq n$. $p = [p_1, p_2, \ldots, p_n]$ is the transmission vector. We assume that the distribution of these packets to the sensors is done in a synchronous time-slotted fashion. In each time slot, an S_i may either receive or transmit (but not both) a packet. To facilitate the transmission of the packets, each S_i has an antenna whose range is r. In the unidirectional antenna model, S_i receives a packet only if that packet is sent in its direction from an antenna located at most r away. Because of interference, a transmission from S_i to S_j is successful iff the following are true:

1. j is in range, that is $d(i, j) \leq r$, where $d(i, j)$ is the distance between S_i and S_j.
2. j is not, itself, transmitting in this time slot.
3. There is no interference from other transmissions in the direction of j. Formally, every S_k, $k \neq i$, that is transmitting in this time slot in the direction of S_j is out of range. Here, out of range means $d(k, j) \geq (1 + \delta)r$, where $\delta > 0$ is an interference constant.

In the omnidirectional antenna model, a packet transmitted by an S_i is received by all S_j (regardless of direction) that are in the antenna's range. The constraints on successful transmission are the same as those for the unidirectional antenna model except that all references to "direction" are dropped. Our objective is to develop an algorithm to complete the specified data distribution using the fewest number of time slots. A related data gathering problem for wireless sensor networks is considered in [58].

2.4   Sensor Fusion

The reliability of a sensor system is enhanced through the use of redundancy. That is, each point or region is monitored by multiple sensors. In a redundant sensor system, we are faced with the problem of fusing or combining the data reported by each of the sensors monitoring a specified point or region. Suppose that $k > 1$ sensors monitor point p. Let m_i, $1 \leq i \leq k$ be the measurement recorded by sensor i for point p. These k measurements may differ because of inherent differences in the k sensors, the relative location of a sensor with respect to p, as well as because one or more sensors is faulty. Let V be the real value for p. The objective of sensor fusion is to take the k measurements, some of which may be faulty, and determine either the correct measurement V or a range in which the correct measurement lies.

The sensor fusion problem is closely related to the Byzantine agreement problem that has been extensively studied in the distributed computing literature [59, 60]. Brooks and Iyengar [61] have proposed a hybrid distributed sensor fusion algorithm that is a combination of the optimal region algorithm of Marzullo [62] and the fast convergence algorithm proposed by Mahaney and Schneider [63] for the inexact-agreement version of the Byzantine generals problem. Let δ_i be the accuracy of sensor i. So, as far as sensor i is concerned, the real value at p is $m_i \pm \delta_i$ (i.e., the value is in the range $[m_i - \delta_i, m_i + \delta_i]$). Each sensor needs to compute a range in which the true value lies as well as the expected true value. For this computation, each sensor sends its m_i and δ_i values to every other sensor. Suppose that sensor i is non-faulty. Then every non-faulty processor receives the correct values of m_i and δ_i. Faulty sensors may receive incorrect values. Similarly, if processor i is faulty, the remaining processors may receive differing values of m_i and δ_i. Algorithm 2.8 gives the Brooks–Iyengar hybrid algorithm [61]. This algorithm is executed by every sensor using as data the measurement ranges received from the remaining sensors monitoring point p plus the sensor's own measurement.

As an example computation, suppose that four sensors S_1, S_2, S_3 and S_4 monitor p and that the four measurement ranges are [2, 6], [3, 8], [4, 10] and [1, 7]. To perform the computation specified by the Brooks–Iyengar hybrid algorithm, the four sensors communicate their measurement ranges to one another. Assume that S_4 is the only faulty sensor. So, sensors S_1, S_2 and S_3 correctly communicate their measurement to one another. However, these sensors may receive differing readings from S_4. Likewise, S_4 may record different receptions from S_1, S_2 and S_3. Let the S_4 measurement received by S_1, S_2 and S_3 be [1, 3], [2, 7], [7, 12], respectively. The V and range computed at each of the four sensors are given in Algorithm 2.9. Note that $k = 4$ and $\tau = 1$.

Algorithm 2.8 Brooks–Iyengar hybrid algorithm to estimate V and range for V

Step 1 (Determine range for real value V)   Let $[l_i, u_i, n_i]$, $1 \leq i \leq q$ be such that

1. $li \leq ui \leq li + 1$, $1 \leq i < q$ and $l_q \leq u_q$. The $[l_i, u_i]$'s define disjoint measurement intervals.
2. $n_i \geq k - \tau$ gives the number of sensors whose measurement range includes $[l_i, u_i]$.
3. If x is a measurement value not included in one of the $[l_i, u_i]$ intervals, x is included in the measurement interval of fewer than $k - \tau$ sensors.

V is estimated to lie in the range $[l_1, u_q]$.

Step 2 (Estimate V) V is estimated to be the weighted average

$$\sum_{i=1}^{q} \frac{(l_i + u_i) \times n_i}{2 \times \sum_{i=1}^{q} n_i}.$$

Algorithm 2.9 Example for Brooks–Iyengar hybrid algorithm

S_1:   The ranges recorded at S_1 are [2, 6], [3, 8], [4, 10], and [1, 3]. There is only
one tuple $[l_i, u_i, n_i]$ that satisfies the Step 1 criteria. This tuple is [4, 6, 3]. S_1
estimates the range for V as [4, 6] and $V = 5$.

S_2:   The recorded ranges are [2, 6], [3, 8], [4, 10], and [2, 7]. The tuples are
[3, 4, 3], [4, 6, 4], and [6, 7, 3]. S_2 estimates the range for V as [3, 7] and $V = 5$.

S_3:   The recorded ranges are [2, 6], [3, 8], [4, 10], and [7, 12]. The tuples are
[4, 6, 3] and [7, 8, 3]. S_3 estimates the range for V as [4, 8] and $V = 6.25$.

S_4:   Since this sensor is faulty, its computation is unreliable. It may compute any
range and value for V.

2.5   Conclusion

We have reviewed some of the recent advances made in the development of
algorithms for wireless sensor networks. This chapter has focussed on sensor
deployment and coverage, routing (specifically, unicast and multicast), and sensor
fusion. Both centralized and distributed localized algorithms have been considered.
In the first section, the algorithms regarding sensor development and coverage were
discussed. Among them, Kar and Banerjee's algorithm was explained in detail.
Also, we discussed about how to evaluate the quality of sensor deployment. In the
second section, we discuss the sensor networks routing algorithms based on unicast,
multicast, and broadcast. Moreover, the algorithms regarding data collection and
distribution were studied. Finally, third section focused on sensor fusion algorithms
including Brooks–Iyengar algorithm.

References

1. R. Szewczyk, E. Osterweil, J. Polastre, M. Hamilton, A. Mainwaring, D. Estrin, Habitat
   monitoring with sensor networks. CACM **47**(6), 34–40 (2004)
2. M.H. Amini, H. Arasteh, P. Siano, Sustainable smart cities through the lens of complex
   interdependent infrastructures: panorama and state-of-the-art, in *Sustainable Interdependent
   Networks*, vol. II. (Springer, Cham, 2019)
3. M.H. Amini, J. Mohammadi, S. Kar. Distributed holistic framework for smart city infrastructures: tale of interdependent electrified transportation network and power grid. IEEE Access **7**,
   157535–157554 (2019)
4. M.H. Amini (ed.), *Optimization, Learning, and Control for Interdependent Complex Networks*.
   Advances in Intelligent Systems and Computing, vol. 2 (Springer, Cham, 2020)
5. A. Imteaj, M.H. Amini, J. Mohammadi. Leveraging decentralized artificial intelligence to
   enhance resilience of energy networks (2019). arXiv preprint:1911.07690
6. S. Iyengar, R. Brooks, Computing and communications in distributed sensor networks. J.
   Parallel Distrib. Comput. **64**(7), 1–1 (2004). Special Issue

7. S. Iyengar, R. Brooks, *Handbook of Distributed Sensor Networks* (Chapman & Hall/CRC, Boca Raton, 2005)
8. D. Culler, W. Hong, Wireless sensor networks. CACM **47**(6), 30 (2004). Special Issue
9. H. Zhang, J. Hou, Maintaining sensing coverage and connectivity in large sensor networks. Technical Report UIUC, UIUCDCS-R-2003-2351 (2003)
10. X. Wang, et al., Integrated coverage and connectivity configuration in wireless sensor networks, in *SenSys* (2003)
11. C. Huang, Y. Tseng, The coverage problem in a wireless sensor network, in *WSNA* (2003)
12. A. Howard, M. Mataric, G. Sukhatme, An incremental self-deployment algorithm for mobile sensor networks. Auton. Robots **13**, 113–126. Special Issue on Intelligent Embedded Systems
13. A. Howard, M. Mataric, G. Sukhatme, Mobile sensor network deployment using potential fields: a distributed, scalable solution to the area coverage problem, in *6th International Symposium on Distributed Autonomous Robotics Systems (DARS02)* (2002)
14. Y. Zou, K. Chakrabarty, Sensor deployment and target localization in distributed sensor networks. ACM Trans. Embed. Comput. Syst. **3**(1), 61–91 (2004)
15. S. Poduri, G. Sukhatme, Constrained coverage for mobile sensor networks, in *IEEE International Conference on Robotics and Automation (ICRA'04)* (2004), pp. 165–171
16. K. Kar, S. Banerjee, Node placement for connected coverage in sensor networks, in *Proceedings of WiOpt 2003: Modeling and Optimization in Mobile, Ad Hoc and Wireless Networks* (2003)
17. K. Chakrabarty, S. Iyengar, H. Qi, E. Cho, Grid coverage for surveillance and target location in distributed sensor networks. IEEE Trans. Comput. **51**(12), 1448–1453 (2002)
18. M. Cardei, D. Du, Improving wireless sensor network lifetime through power aware organization. ACM Wirel. Netw.; *IEEE International Conference on Communications (ICC 2001)*, Helsinki (11–14 June 2001, to appear)
19. S. Slijepcevic, M. Potkonjak, Power efficient optimization of wireless sensor networks, in *IEEE International Conference on Communications* (2001)
20. F. Ye, G. Zhong, S. Lu, L. Zhang, Energy efficient robust sensing coverage in large sensor networks. Technical Report, UCLA, 2002
21. Y. Xu, J. Heidermann, D. Estrin, Geography-informed energy conservation for ad hoc routing, in *MOBICOM* (2001)
22. D. Tian, N.D. Georganas, A coverage-preserving node scheduling scheme for large wireless sensor networks, in *Proceedings of the 1st ACM international workshop on Wireless sensor networks and applications* (ACM, New York, 2002), pp. 32–41
23. F. Ye, G. Zhong, S. Lu, L. Zhang, Peas: a robust energy conserving protocol for long-lived sensor networks, in *23rd ICDCS* (2003)
24. T. Yan, T. He, J. Stankovic, Differentiated surveillance for sensor networks, in *First International Conference on Embedded Networked Sensor Systems* (2003), pp. 51–62
25. J. Lu, S. Tatsuya, Coverage-aware self-scheduling in sensor networks, in *IEEE Computer Communications Workshop (CCW 2003)* (2003)
26. S. Meguerdichian, F. Koushanfar, M. Potkonjak, M. Srivastava, Coverage problems in wireless ad-hoc sensor networks, in *IEEE InfoCom* (2001)
27. X. Li, P. Wan, O. Frieder, Coverage in wireless ad-hoc sensor networks. IEEE Trans. Comput. **52**, 753–763 (2002)
28. S. Meguerdichian, F. Koushanfar, G. Qu, M. Potkonjak, Exposure in wireless ad hoc sensor networks, in *7th Annual International Conference on Mobile Computing and Networking (MobiCom'01)* (2001), pp. 139–150
29. G. Veltri, Q. Huang, G. Qu, M. Potkonjak, Minimal and maximal exposure path algorithms for wireless embedded sensor networks, in *SenSys'03* (2003)
30. R. Kannan, S. Sarangi, S. Ray, S. Iyengar, Minimal sensor integrity: computing the vulnerability of sensor grids. Info. Proc. Lett. **86**(1), 49–55 (2003) 26
31. R. Kannan, S.Sarangi, S.S. Iyengar, L. Ray, Sensor-centric quality of routing in sensor networks, in *INFOCOM* (2003)

32. R. Kannan, S.S. Iyengar, Game-theoretic models for reliable, path-length and energy-constrained routing in wireless sensor networks. IEEE J. Sel. Areas Commun. **22**, 1141–1150 (2004)
33. T. Rappaport, *Wireless Communications: Principles and Practices* (Prentice Hall, New Jersey, 1996)
34. S. Singh, M. Woo, C. Raghavendra, Power-aware routing in mobile ad hoc networks, in *ACM/IEEE MOBICOM* (1998)
35. S. Sahni, *Data Structures, Algorithms, and Applications in Java*, 2nd edn. (Silicon Press, New Jersey, 2005)
36. J. Aslam, Q. Li, R. Rus, Three power-aware routing algorithms for sensor network. Wirel. Commun. Mobile Comput. **3**, 187–208 (2003)
37. K. Kar, M. Kodialam, T. Lakshman, L. Tassiulas, Routing for network capacity maximization in energy-constrained ad-hoc networks, in *IEEE INFOCOM* (2003)
38. C. Florens, R. McEliece, Scheduling algorithms for wireless ad-hoc sensor networks, in *IEEE GLOBECOM* (2002), pp. 6–10
39. A. Misra, S. Banerjee, MRPC: maximizing network lifetime for reliable routing in wireless, in *IEEE Wireless Communications and Networking Conference (WCNC)* (2002)
40. J. Chang, L. Tassiulas, Routing for maximum system lifetime in wireless ad-hoc networks, in *37th Annual Allerton Conference on Communication, Control, and Computing, Monticello, IL, September* (1999)
41. J. Chang, L. Tassiluas, Energy conserving routing in wireless ad-hoc networks, in *IEEE INFOCOM* (2000)
42. J. Wu, M. Gao, I. Strojmenovic, On calculating power-aware connected dominating sets for efficient routing in ad hoc wireless networks. J. Commun. Netw. **4**(1), 59–70 (2002)
43. I. Stojmenovic, X. Lin, Power-aware localized routing in wireless networks. IEEE Trans. Parallel Distrib. Syst. **12**, 1122–1133 (2000)
44. T. Melodia, D. Pompili, I. Akyildiz, Optimal local topology knowledge for energy efficient geographical routing in sensor networks, in *IEEE INFOCOM* (2004)
45. W. Heinzelman, A. Chandrakasan, H. Balakrishnan, Energy-efficient communication protocol for wireless microsensor networks, in *IEEE HICSS* (2000)
46. J. Wieselthier, G. Nguyen, A. Ephremides, On the construction of energy-efficient broadcasting and multicast trees in wireless networks, in *IEEE INFOCOM* (2000)
47. J. Wieselthier, G. Nguyen, Algorithm for energy-efficient multicasting in static ad hoc wireless networks. Mobile Netw. Appl. **6**, 251–261 (2001)
48. M. Garey, D. Johnson, *Computers and Intractability: A Guide to the Theory of NP-completeness* (W. H. Freeman, New York, 1979)
49. S. Guha, S. Khuller, Approximation algorithms for connected dominating sets, in *Fourth Annual European Symposium on Algorithms* (1996)
50. R. Marks, A. Das, M. El-Sharkawi, P. Arabshahi, A. Gray, Minimum power broadcast trees for wireless networks: optimizing using the viability lemma, in *IEEE ISCAS* (2002)
51. A.K. Das, R.J. Marks, M. El-Sharkawi, Minimum power broadcast trees for wireless networks, in *IEEE International Symposium on Circuits and Systems, May* (2002)
52. P. Wan, G. Calinescu, X. Li, O. Frieder, Minimum-energy broadcast routing in static ad hoc wireless networks, in *IEEE INFOCOM* (2001)
53. J. Park, S. Sahni, Maximum lifetime routing in wireless sensor networks. In *Computer and Information Science and Engineering, University of Florida* (2005)
54. J. Cartigny, D. Simplot, I. Stojmenovic, Localized minimum-energy broadcasting in ad-hoc networks, in *IEEE INFOCOM* (2003)
55. J. Wu, F. Dai, Broadcasting in ad-hoc networks based on self-pruning, in *IEEE INFOCOM* (2003)
56. S. Singh, C. Raghavendra, J. Stepanek, Power-aware broadcasting in mobile ad hoc networks, in *IEEE PIMRC'99, Osaka, Japan* (September 1999)
57. C. Florens, R. McEliece, Packets distribution algorithms for sensor networks, in *INFOCOM* (2003)

58. Y. Yu, B. Krishnamachari, V. Prasanna, Energy-latency tradeoffs for data gathering in wireless sensor networks, in *INFOCOM* (2004)
59. L. Lamport, R. Shostak, M. Pease, The Byzantine generals problem. ACM Trans. Program. Lang. Syst., NA, 499–516 (1986)
60. D. Dolev, The Byzantine generals strike again. J. Algorithms **3**, 14–30 (1982)
61. R. Brooks, S. Iyengar, Robust distributed computing and sensing algorithm. IEEE Comput. **29**, 53–60 (1996)
62. K. Marzullo, Tolerating failures of continuous-valued sensors. ACM Trans. Comput. Syst. **8**, 284–304 (1990)
63. S. Mahaney, F. Schneider, Inexact agreement: accuracy, precision, and graceful degradation, in *Fourth ACM Symposium on Principles of Distributed Computing* (1985), pp. 237–249

Chapter 3
Fault Tolerant Distributed Sensor Networks

3.1 Introduction

Our modern world contains many automated systems that must interact with changing environments. This increases complexity in complex networks and requires
developing efficient computational algorithms [1], such as distributed optimization
methods for large-scale problems in interdependent networks [2], federated learning
for IoT devices [3], and multi-agent systems for energy systems [4]. Because these
environments cannot be predetermined, the systems rely on sensors to provide them
with the information they need to perform their tasks. Sensors providing data for
control systems are the unenviable interface between computer systems and the real
world. Programming automated control systems is difficult because sensors have
limited accuracy, and the readings they return are frequently corrupted by noise.

To avoid systems being vulnerable to a single component failure, it is reasonable
to use several sensors redundantly. For example, an automatic tracking system could
use different kinds of sensors (radar, infrared, microwave) that are not vulnerable
to the same kinds of interference. Redundancy presents a new problem to system
designers because the system will receive several readings that are either partially
or entirely in error. It must decide which components are faulty, as well as how to
interpret at least partially contradictory readings.

To improve sensor system reliability, researchers have actively studied the
practical problem of combining, or fusing, the data from many independent sensors
into one reliable sensor reading. When integrating sensor readings, robustness and

---

In preparation of this chapter, the authors reused some parts of the following article with
permission: "Brooks, Richard R., and S. Sitharama Iyengar. Robust distributed computing and
sensing algorithm. Computer 29.6 (1996): 53–60."

© Springer Nature Switzerland AG 2020
P. Sniatala et al., *Fundamentals of Brooks–Iyengar Distributed Sensing Algorithm*,
https://doi.org/10.1007/978-3-030-33132-0_3

reliability are crucial properties. It is increasingly obvious that sensor integration, which must include some type of fusion, is necessary to automate numerous critical systems.

Redundant sensors in an automated control system form one type of distributed system. A key advantage of distributed computing is that it adds a new dimension of integrity to computing. Computations made by a network of independent processors are insensitive to a single hardware failure. Instead, the concerns in a distributed system are determining how many component failures a network can tolerate, how the network separates the output from correctly functioning and still be reliable, and how the network separates the output from correctly functioning machines from that of defective machines.

The central question is, how can an automated system be certain to make the correct decision in the presence of faulty data? Much depends on the system's accuracy—the distance between its results and the desired results—and on the system's precision—the size of the value range it returns.

To solve the problem algorithmically, we basically have sensor fusion and Byzantine agreement. Dolev et al. [5] presented one Byzantine agreement algorithm to solve the Byzantine generals problem posed by Lamport and colleague [6]. The Byzantine generals problem presupposes a distributed decision-making process in which some participants not only make the wrong decision but maliciously attempt to force disagreement within the group. An algorithm that solves this problem can reliably be used in distributed computing, because even the failure of a limited number of machines in a network cannot cause the network to malfunction.

In this chapter, we describe a hybrid algorithm we developed that satisfies both the precision and accuracy requirements of distributed systems. We used established methods for distributed agreement based on data of limited accuracy. The inexact-agreement and approximate agreement algorithms have been successfully used for implementing clock synchronization protocols and proposed as models for sensor averaging [5, 7]. The sensor fusion algorithm is well established as a method for accurately averaging sensor readings [8, 9]. Our hybrid algorithm is suitable for use in both environments and manages to provide increased precision for distributed decision-making without adversely affecting system accuracy.

3.2 Byzantine Generals Problem

The Byzantine generals problem concerns a mythical siege of a city by the Byzantine army [10]. The army's commander-in-chief has several troops positioned around the city. Each position is under a general's command. The commander-in-chief knows that many of his generals and messengers are traitors who are loyal to the opposing army. He must tell all his generals to either attack or retreat.

The generals can discuss the decision among themselves via messengers. If all loyal armies follow the orders of a loyal commander-in-chief, they stand a good chance of success. If one part of the army attacks while another part retreats, they face certain defeat. How can the loyal generals guarantee, by exchanging messages among themselves, that all generals make the same decision, and that this decision is given by a loyal commander-in-chief [6].

This problem is directly applicable to distributed computing. It can be rephrased as a system of N independent processing elements (PEs), up to τ of which may be faulty. We must develop a protocol that guarantees for all messages broadcast by any processor X:

- The non-faulty processors agree among each other on contents of the data received from X.
- If X is non-faulty, the agreement should equal the contents of the message sent from X.

This is also called *general interactive consistency* [11].

This problem has some interesting characteristic [6]. It can be solved only if τ, the number of traitors, is less than one-third of N, the total number of PEs. The proof in Fig. 3.1 is done by showing that, in a graph of only three nodes with one faulty node, it is impossible for a correct PE to determine which of the other two nodes is faulty.

Dolev showed that τ must be less than half the connectivity of the graph [12]. This is intuitively evident, as Fig. 3.1 shows: Because a node can change messages passing through it, any part of the graph that receives a majority of messages potentially modified by traitors will be deceived. In other words, to tolerate τ faults, the system must have at least $\tau + 1$ PEs, and every PE must be connected directly to at least $\tau + 1$ other PEs [12]. It has been proven as well that an algorithm that solves the Byzantine generals problem must execute at least $\tau + 1$ rounds of broadcasts between nodes [13].

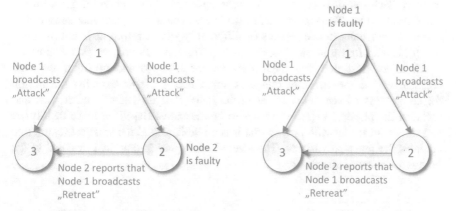

Fig. 3.1 Node 3 cannot distinguish between the two scenarios shown. It is impossible for it to determine if node 1 or node 2 is the faulty node

Many algorithms have been found to solve the Byzantine generals problem. Dolev presented a typical one [12]. We will not present the details of these algorithms here, but it is worth noting that they require each node to rebroadcast all the information it has received. A lower bound of $O(N\tau)$ messages must be broadcast to ensure agreement, and the message size grows exponentially with each round, finishing with size $O(N\tau + 1)$ [10].

3.3   Fault Tolerant Sensor Fusion

Sensors play a pivotal role to in networks which their performances rely on the sensor measurements to operate properly and perform tasks in time-varying environments. However, due to the measurement errors of these sensors, uncertainties, and physical failures, the network performance is vulnerable to their accurate operation. In order to make the system vulnerable to a single component failure, several sensors are deployed to increase the redundancy which further increases the reliability. For instance, the same solution is applied to automatic tracking systems. Although redundancy may help to deal with the vulnerability concerns, it brings a new problem because of the inevitable erroneous readings.

Marzullo [8] introduced a new model with abstract sensors[1] and concrete sensors.[2] Due to the limited accuracy of all sensors, the communicated values by an abstract sensor includes a lower bound and an upper bound. In this vein, sensor inaccuracies can be dealt with explicitly.

We address the following sensor fusion problem: *Given a set of N sensors with a limited accuracy and at most τ of the sensors being faulty, what is the smallest value range where we can be certain to find the correct value?*

One of the proposed algorithms is approximate-matching algorithm. Let N denote the set of real values whose accuracy is defined by being within a specific distance of the unique correct value, as well as up to τ values that may be arbitrarily in error. Each sensor is represented by the upper and lower bound it gives for the value it measures. The size of this range is the accuracy, [8], for that sensor. It is often advantageous to use sensors of different types; therefore, it is not assumed that [8] is uniform for all sensors, and the algorithm makes no precision restraints.

Marzullo illustrated that measurements are effective and useful if they are accurate and if the specified range is small enough. This explains the reason for correctness of sensors only when the range it returns is of limited size and includes the physical expected values to be measures. Based on these definitions and restrictions, Marzullo proved that it is possible to find the smallest region that contains the physical variable. The size of this region is less than or equal to the largest accuracy of all sensors.

---

[1] An abstract sensor consists of a range of values measured and communicated by a sensor.

[2] A concrete sensor is the physical device that returns the value.

Finding all regions in which $N - \tau$ sensor readings intersect provides a correct solution to the sensor fusion problem. It is easy to see that the correct value must be in one of these intervals. The correct range is thus defined by the value of the smallest lower bound and the largest upper bound of these intersections [8].

Our optimal region algorithm modifies Brooks' and Iyengar's [14] multidimensional algorithm and is equivalent to Marzullo's algorithm:

Algorithm 3.1: OPTIMALREGION

Input:
    A set of sensor measurements S.
Output:
    The feasible region described by the correct space.

1: Initialize a set of regions, denoted by C, to NULL.
2: Sort all points of S in an ascending fashion
3: A reading is considered active if its lower bound has been sensed but its upper bound has yet to be sensed. If a region is reached the status in which more than $N - \tau$ readings are active, add the region to C.
4: All points have been processed. List C now contains intersections of at least $N - \tau$ sensor readings. Sort the intersections in C.
5: Output the region defined by the minimum lower bound and the largest upper bound in C.

Algorithm 3.1 specifies the five-step procedure to determine the feasible region described in the correct space given the set of sensor measurements denoted by S.

This algorithm can be implemented for a distributed sensor network, where all sensors (PEs) analyze information in parallel. In a distributed environment, the algorithm's precision is bounded by the accuracy.

3.3.1   Precision and Accuracy

Two types of methods accept data from several sources, where a minority of the data is intentionally incorrect, and the correct part of data is incomplete: (1) approximate agreement algorithm, (2) sensor fusion algorithm. These algorithms try to find a result that accurately represents the data from the correct PEs, while minimizing the influence of data from faulty PEs.

Approximate agreement emphasizes precision, even when this conflicts with system accuracy. For both approximate and inexact agreement, the overall accuracy of the readings remains constant and can decrease for some PEs. This approach is justified when processor coordination is more important than the quality of the results. For example, these algorithms are useful for clock synchronization among PEs.

Sensor fusion, on the other hand, is mainly concerned with the accuracy of the readings, which is appropriate for sensor applications. This is true despite the fact that increased precision within known accuracy bounds would be beneficial for coordinating distributed automation systems.

Any desired level of precision can be attained by iterating the approximate agreement algorithm. The accuracy requirements of inexact agreement could eventually make it impossible to perform further iterations, putting a lower limit on the achievable precision. It is not feasible to achieve a higher level of accuracy by repeating the sensor fusion algorithms.

By allowing each PE to have its specific accuracy, the sensor fusion algorithm allows the use of heterogeneous sensors. The bandwidth needed for one iteration of all algorithms not containing Byzantine agreement is linear in the number of processors. If a Byzantine pre-processing step is added [7, 15], the number of communicated bits will be exponential in the number of rounds R needed for Byzantine agreement on the values, $O(NR + 1)$ where R will be greater than $\tau + 1$ [15]. Some algorithms perform agreement with a polynomial order but more required iterations, which leads making performance less than asymptotically optimal. Consequently, algorithms with a Byzantine agreement step are not proper for real-time systems.

The complexity of the sensor fusion and FCA algorithms is $O(N \log(N))$. It is worth noting that Mahaney and Schneider's "acceptability" criteria correspond to PEs whose values intersect with the values of at least $N - \tau$ other PEs. Marzullo's and the sensor fusion algorithms can be improved to determine whether a PE is "acceptable."

One iteration of Dolev's approximate agreement algorithm requires finding the τ largest and τ smallest of N values. To this end, one can find and remove the maximum and minimum boundaries of N values for τ times. Alternatively, one can sort the elements of this set in increasing order. The algorithm can therefore be implemented in $\min(O(N\tau), O(N \log N))$ time, assuming the following inequality holds: $\tau < \log N$.

3.3.2 Brooks–Iyengar Algorithm

In order to meet the needs of both the inexact-agreement and the sensor fusion problems, we integrated the optimal region algorithm with FCA to generate an algorithm that achieve the best accuracy possible and enhances the precision of distributed decision-making.

The data used by this algorithm can take any of the following three forms:

- The values of all real parameters have the same implicit accuracy.
- If the value of a real parameter has explicit accuracy, it will be transmitted along with the value.
- A range consisting of an upper and lower bound.

The asymptotic complexity of this algorithm, like FCA and the sensor fusion algorithm, is $O(N \log N)$. This is due to step 2, which requires sorting the input based on the lower bound values. Steps 3 and 4 are specifically for the ranges where $N - \tau$ or more values intersect. Since there are at most $2\tau + 1$ such ranges [14], step 3 has complexity $O(\tau \log \tau)$, and step 4 has $O(\tau)$, both of which are less than $(N \log N)$.

Algorithm 3.2: Brooks-Iyengar hybrid

Input:
  A set of data S.
Output:

  A real number leads to accurate solution and a range giving its explicit accuracy bounds.
1: Each PE receives the values from all other PEs and forms a set V.
2: Perform the optimal region algorithm on V and return a set A that consists of a range for
   values where at least $N - \tau$ PEs intersect. a
3: Output the range defined by the lowest lower bound and the largest upper bound in A.
   These values specify the accuracy bounds of the answer.
4: Sum the the midpoints of each range in A is multiplied by the number of sensors whose
   readings are a subset of the specified range, and divide by the number of factors. This is the
   answer.

Algorithm 3.2 consists of four consecutive steps through which it calculates the accurate solution as a real number and a range giving its explicit accuracy bounds for a given data set S.

In the Marzullo's sensor fusion approach, the hybrid algorithm deals with all data as a range of possible values rather than a discrete point value. Since the problem relates data with accuracy and precision limitations, computations based on point values can be misleading. This difference is useful conceptually and can also be advantageous in increasing the precision and accuracy of the results.

Our algorithm's accuracy is identical to that of the optimal region algorithm. Since the real value computed in step 4 is bounded by this range, it is within the accuracy bounds defined by Mahaney and Schneider [7] and by Marzullo [8]. The accuracy bounds returned by the algorithm maintain the Dolev's accuracy constraint as a subset.

The precision analysis Mahaney and Schneider performed is valid for the hybrid algorithm when uniform accuracy is assumed. The values will converge, despite arbitrary errors, at a rate of at least $2\tau/N$ per iteration [7]. This convergence must be considered as a lower bound, since the worst-case scenario of this algorithm is identical to Mahaney and Schneider's FCA algorithm. If uniform accuracy is not assumed, it is impossible to define bounds on the convergence rate, short of it being less than 1, since we are dealing with a set of arbitrarily distributed numbers.

In many cases, the hybrid algorithm outperforms the FCA. FCA computes its value based on a simple function of all values that satisfy the acceptability requirement, i.e., the values should be in distance δ of $N - \tau - 1$. The values the faulty PE which are sent to other nodes may have different values by up to 4δ. The different values are then entered directly into the function $e(A)$, which computes the value at each node.

The hybrid algorithm uses a different approach than FCA. It is based on the midpoint of the region where the erroneous measurements intersect with at least $N - \tau - 1$ correct readings. Using the example of a system with four PEs, the value transmitted to two different nodes by the faulty PE can still vary by up to 4δ and be acceptable. The value used in calculations by the hybrid algorithm, however, differs by at most 2δ.

The hybrid algorithm also weights each region by the number of PEs that belong to a specific region. This increases accuracy by enhancing the influence of PEs that cover more than one set of PEs. A faulty answer is most disruptive when it barely fulfills the acceptability requirement and forms one intersection with $N - \tau - 1$ other PEs. When a PE's value contributes to many acceptable regions, it is closer to the value that the hybrid algorithm will calculate. Weighting the region midpoints exploits this fact to increase the algorithm's convergence.

It is also be possible to insert another step between steps 1 and 2 that performs Byzantine agreement among the PEs. This would correspond to Mahaney and Schneider's CCA algorithm.

3.3.3 Where Does Brooks–Iyengar Algorithm Stand?

Figure 3.2 illustrates the key role of Brooks–Iyengar algorithm in evolving the literature of sensing algorithms:

- 1982 Byzantine Problem: The Byzantine General Problem as an extension of Two Generals' Problem could be viewed as a binary problem.
- 1983 Approximate Consensus: The method removes some values from the set consists of scalars to tolerant faulty inputs.
- 1985 Inexact Consensus: The method also uses scalar as the input.
- 1996 Brooks–Iyengar algorithm: The method is based on intervals.
- 2013 Byzantine Vector Consensus: The method uses vectors as the input.
- 2013 Multidimensional Agreement: The method also use vectors as the input while the measure of distance is different.
- We could use Approximate Consensus (scalar-based), Brooks–Iyengar algorithm (interval-based), and Byzantine Vector Consensus (vector-based) to deal with interval inputs, and the paper proved that Brooks–Iyengar algorithm is a valid solution in this problem.

Algorithm Characteristics
- Faulty PEs tolerated $< \frac{N}{3}$
- Maximum faulty PEs $< \frac{2N}{3}$

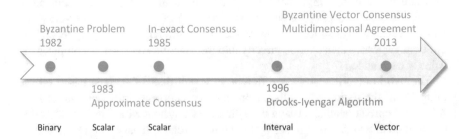

Fig. 3.2 Brooks–Iyengar algorithm among others

- Complexity $= O(N \log N)$
- Order of network bandwidth $= O(N)$
- Convergence $= \frac{2t}{N}$
- Accuracy = limited by input
- Iterates for precision = often
- Precision over accuracy = no
- Accuracy over precision = no

3.3.4   Comparing the Performance of Different Algorithms

Figure 3.3 illustrates a scenario that highlights the differences among the above-mentioned algorithms. Multiple ground radar stations are detecting a target's position.

We have $N = 5$, and therefore τ cannot be greater than 1. Four sites, $S1$, $S2$, $S3$, and $S4$, function correctly. The fifth site, $S5$, is faulty; it broadcasts different values to each of the other four sites. Table 3.1 shows the values.

The four correct sites agree on the range [2.7, 2.8]. So the actual value must lie within this range.

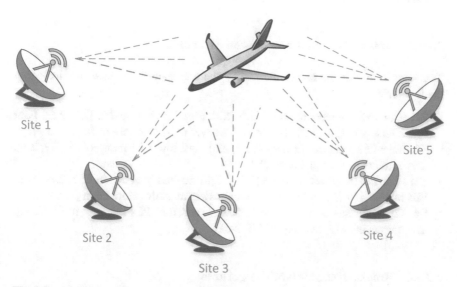

Site 1

Site 5

Site 2

Site 4

Site 3

Fig. 3.3 Multiple ground radar stations are detecting a target's position. To coordinate actions, you need a single accurate reading

Table 3.1 $S5$ values broadcast to other sites

–	$S1$	$S2$	$S3$	$S4$
Value	[2.7, 6.7]	[0, 3.2]	[1.5, 4.5]	[0.8, 2.8]
$S5$ value	[1.4, 4.6]	[−0.6, 2.6]	[0.9, 4.1]	[−0.7, 2.5]

3.3.4.1   Dolev's Algorithm

Since τ is 1, Dolev's approximate agreement algorithm discards the highest and lowest value at each PE and averages the remaining values.

- **S1:** Discard 1.6 and 4.7. Averaging the remaining values gives an answer of 2.6.
- **S2:** Discard 1.0 and 4.7. Averaging the remaining values gives 2.13.
- **S3:** Discard 1.6 and 4.7. Averaging the remaining values gives 2.43.
- **S4:** Discard 0.9 and 4.7. The problem becomes identical to the one for S2, and the answer is again 2.13.

3.3.4.2   Mahaney and Schneider's FCA Algorithm

All values sent by $S5$ are within their accuracy range of an intersection with at least three other sites. The values from $S5$ will not be disqualified for use by any site. The answer for each site is a simple average of all five values present at that site.

- **PE S1:** Calculates a value of 2.82.
- **PE S2:** Calculates 2.42.
- **PE S3:** Calculates 2.72.
- **PE S4:** Calculates 2.4.

3.3.4.3   Optimal Region Sensor Fusion Algorithm

This algorithm uses the ranges defined by the uncertainties of each site, including the faulty $S5$.

- **S1:** Four PEs intersect in range $[1.5, 2.7]$. Five PEs intersect in $[2.7, 2.8]$. Four PEs intersect in $[2.8, 3.2]$. The correct answer must therefore lie in $[1.5, 3.2]$.
- **S2:** Four PEs intersect in range $[1.5, 2.6]$, and four PEs intersect in $[2.7, 2.8]$. The correct answer must be in $[1.5, 2.8]$.
- **S3:** Four PEs intersect in range $[1.5, 2.7]$. Five PEs intersect in $[2.7, 2.8]$. Four PEs intersect in $[2.8, 3.2]$. The correct answer must therefore lie in $[1.5, 3.2]$.
- **S4:** Four PEs intersect in range $[1.5, 2.5]$, and four PEs intersect in $[2.7, 2.8]$. The correct answer must be in $[1.5, 2.81]$.

3.3.4.4   Brooks–Iyengar Hybrid Algorithm

The obtained ranges by the sensor fusion algorithm are the accuracy limits that hybrid algorithm returns for each site. In order to determine the answer for a site, the hybrid algorithm makes a weighted average of the midpoints of the regions found by the sensor fusion algorithm.

- **S1:** The weighted average is: $(4*2.1 + 5*2.75 + 4*3.0)/13 = 2.625$.
- **S2:** The weighted average is: $(4*2.05 + 4*2.75)/8 = 2.4$.

- **S3:** The weighted average is: $(4*2.1 + 5*2.75 + 4*3.0)/13 = 2.625$.
- **S4:** The weighted average is: $(4*2.0 + 4*2.75)/8 = 2.375$.

3.3.4.5   Results

All four algorithms are designed to find the results within the accuracy and precision bounds that are specified by the algorithm designers. All three algorithms designed for improving precision Dolev's, Mahaney and Schneider's returned answers within a narrower range than the input data. For the illustrative example, the hybrid algorithm's output was the most precise answer. It is worth mentioning that all solutions were lying within a range of length 0.25.

Brooks–Iyengar's hybrid algorithm effectively solves the problem of making the correct decision in the presence of faulty data. This is important for several reasons:

- Because the same algorithm can enhance both accuracy and precision, many real-world distributed applications can use one unified general algorithm.
- The derivations of sensor fusion and approximate agreement are independent. One is based on set theory; the other, on geometry. It thus produces two explanations of the same problem, which facilitate the solution.
- The hybrid algorithm can be extended to solve problems in many applications.

Use of this algorithm is not limited to the mentioned applications. All floating-point computations have limited accuracy that differs from machine to machine. The hybrid algorithm affords scientific computing increased reliability by letting calculations be performed on a distributed system comprising heterogeneous components. This provides a method which is more resistant to round-off and to skewing errors resulting from hardware limitations.

Software reliability is a growing concern. Programs written independently for different hardware platforms that produce equivalent output create a situation like the tracking system example with multiple sensor input. It is a valid assumption that many independent programming teams would not make exactly the same programming error, and that of N such programs fewer than one-third is incorrect for a given instance.

This is analogous to using many sensors based on different technologies—thus sensitive to different types of noise and interference—to measure the same physical entity. The failings of one hardware component will be compensated for by the use of another component made in a complementary manner. These assumptions, used to justify N-modular redundancy for hardware systems, could be valid for software.

Critical modules of important systems require extra effort to ensure their reliability. Reliability requires both precision and accuracy. Unfortunately, there is often a trade-off to be made between the two. The hybrid algorithm we have described allows for increased precision, without sacrificing accuracy in the process.

The algorithm lets distributed systems converge toward an answer that lies within precisely defined accuracy bounds. With this algorithm, truly robust distributed computing applications can be developed.

3.4   Theoretical Analysis of Distributed Agreement Algorithms

Brooks–Iyengar hybrid algorithm is a distributed algorithm that improves both the precision and accuracy of the interval measurements by a distributed sensor network. This algorithm works even in the presence of faulty sensors. The sensor network performs this task by exchanging the measured and accuracy values at every node with the rest of nodes in the network. It further computes the accuracy range and a measured value for the whole network based on collected values.

In order to enable distributed control in the presence of noisy data, the Brooks–Iyengar hybrid algorithm combines Byzantine agreement with sensor fusion. It bridges the gap between sensor fusion and Byzantine fault tolerance. This seminal algorithm unified these disparate fields for the first time. It combines Dolev's algorithm for approximate agreement with Mahaney and Schneider's fast convergence algorithm (FCA). The algorithm assumes N processing elements (PEs), t of which are faulty and can behave maliciously. It takes as input either real values with inherent inaccuracy or noise (which can be unknown), or a real value with *apriori* defined uncertainty, or an interval. The output of the algorithm is a real value with an explicitly specified accuracy. The algorithm runs in $O(N\log N)$ where N is the number of PEs. This algorithm has several real-world applications, including distributed control, software reliability, and high-performance computing.

3.4.1   Background

Consensus in distributed systems and sensor fusion are fundamental problems. Applications of these problems exist in time synchronization, sensor networks, and other domains [10, 16–18]. Fault tolerance is a crucial challenge [19–21] and several algorithms have been introduced for distributed sensing, information fusion, and time synchronization [10, 16, 20, 22–24].

To formulate this problem, a network of N PEs is considered. Note that $\tau < N$ of the PEs may be faulty and provide erroneous data. To address this issues, the faulty PEs are assumed to be maliciously conspire to create the worst possible set of inputs to force the algorithm to fail. If the approach can function appropriately, independent from the set of inputs, then it effectively solves the problem. Concentrating on worst-case analysis can also establish effective performance bounds.

Consider a set of distributed PE's, each PE_i $(i \leq N)$ measures some parameter with a noise factor that introduces a random, bounded deviation from the correct value. An unknown subset of the PE's are faulty. A faulty PE can send an arbitrary measurement to each collaborating PE. Ignoring the network topology, for the system of N PE's to be able to reach consensus, the number of false inputs must be less than $N/3$. A proof of this limit is in [6].

We now review the set of fusion approaches most relevant to this problem:

- *Although the Byzantine General Problem (BGP)* for synchronous systems was tackled in [25] and [6], Fischer [26] proved that in asynchronous systems it is generally not possible to guarantee the convergence. In [27], it has been proved that the convergence is possible in partially asynchronous systems, with the presence of some synchrony. *Dolev et al's approximate agreement* algorithm achieved consensus within known precision bounds. Fekete [15] improved Dolev's algorithm to ameliorate the convergence.
- *Mahaney and Schneider's inexact agreement* took into account both accuracy and precision. Abraham [28] presented a deterministic optimal resilience approximate agreement algorithm that can tolerate more Byzantine faults than Dolev's work [5]. More recently, Vaidya studied the Iterative Approximate Byzantine Consensus (IABC) algorithm to reach consensus for an arbitrary directed graph [29–31]. Approximate agreement has also been studied in dynamic networks [32–34].
- *Byzantine Vector Consensus (BVC)* reaches consensus when inputs are vectors under complete graphs [35] and incomplete graphs [36]. Mendes et al. [37] studied the multidimensional approximate agreement in asynchronous systems, where the *agreement* definition is different: the distance used in multidimensional approximate agreement is Euclidean distance. Another form of consensus in Byzantine asynchronous system is to agree on a vector, while the PEs' input is scalar [38, 39].
- *Marzullo [8]* proposed a fault tolerance fusion approach that deploys interval inputs that finds an interval where all valid readings intersect. In most cases, Marzullo's approach achieved better accuracy than individual sensor inputs. The fused interval is at least as accurate as the range of the least accurate individual sensor. Marzullo's algorithm has been used for clock synchronization [40, 41] and information integration [42–44]. Blum [45] found the worst-case for Marzullo's method in clock synchronization. To reduce the output interval width of Marzullo's method, Prasad's method chose the interval where most input intervals intersect [46, 47]. In order to satisfy the Lipschitz conditions that minor changes of input cause only negligible changes of output [48], Schmid proposed [49] that the output interval might be larger than Marzullo's method. Jayasimha [50] extended the original Marzullo's method to identify sensors and combinations of sensors that are faulty. Chew and Marzullo extended the original Marzullo's approach from one dimension to multi-dimension [9]. Besides Marzullo's algorithm and its related studies, Desovski [51] developed a randomized voting algorithm to select the interval that has votes larger than v, where v is determined by numbers of faulty sensors. Parhami [52] considered interval voting that combines either preference or uncertainty intervals.

- *Multi-Resolution Decomposition* techniques have been applied to fault tolerance. Prasad [53, 54] employed Multi-Resolution Integration (MRI) to recognize and isolate the most prominent and robust peaks in a region for fault tolerant sensor integration when the numbers of sensors are large and large portion of sensor faults are tame. Qi [55] built on the original MRI for mobile-agent-based distributed sensor network.
- *The Brooks–Iyengar algorithm* [56, 57] was used as a distributed tracking algorithm in the DARPA Sense-IT program and was then applied to a real-time extension of Linux [58]. This algorithm considers intervals where $N - \tau$ intervals overlap and performs a weighted average of the interval midpoints. This minimizes the influence of faulty inputs by only considering ranges where faulty inputs agree with a number of valid inputs.

Instead of assuming the presence of malicious inputs, other fusion methods assume data is contaminated by a limited amount of noise [59–61] that is Gaussian. Fusion utilizes tools from probability for instance maximum likelihood estimation. The PEs try to agree on a value only by iteratively exchanging information with neighbors. Sensor fusion by Dempster–Shafer Theory in areas of autonomous mobile robots, context sensing, and target identification has also been proposed [62–64].

In the rest of this section, the objective of distributed sensor fusion is to achieve consensus among the PEs and minimize the impact of faulty data. This section establishes precision bounds for five fault tolerant fusion algorithms that are representative of larger classes. The notation used in this section is in Table 3.2.

Table 3.2 List of notation

Notation	Description
$\widehat{v}$	The ground truth value being measured
N	Total number of PEs
τ	The number of faulty PEs
$v_{j \to i}$	Measurement received by PE_i from PE_j
V_i	The multiset of measurements received by PE_i, i.e., $\{v_{1 \to i}, \ldots, v_{N \to i}\}$
g	The multiset of measurements from non-faulty/valid PEs
G	The set of all possible valid measurements, so $g \in G$
$\tilde{v}_i$	Fused output at PE_i
$\delta(g)$	The maximum difference between any two non-faulty measurements, i.e., $\max(g) - \min(g)$
ε	Precision of fusion results
ζ	Accuracy of fusion results
$[l_{j \to i}, h_{j \to i}]$	The interval measurement sent to PE_i by PE_j

3.4.2 Naive Averaging

We start by considering one of the simplest fusion approaches. Consider a network of N PEs. Each PE broadcasts its local measurement to all the other PEs and calculates its output as the average value of the measurements it receives. Let $v_{j \to i}$ denote the measurement received by PE_i from PE_j, the output $\tilde{v}_i$ using naive averaging is

$$\tilde{v}_i = \frac{v_{1 \to i} + v_{2 \to i} + \cdots + v_{N \to i}}{N}$$

If we assume PE_k sends different readings to other PEs, the difference between the outputs at PE_i and PE_j is

$$
\begin{aligned}
&\left| \tilde{v}_i - \tilde{v}_j \right| \\
&= \left| \frac{v_{1 \to i} + \cdots + v_{k \to i} + \cdots + v_{N \to i}}{N} - \frac{v_{1 \to j} + \cdots + v_{k \to j} + \cdots + v_{N \to j}}{N} \right| \\
&= \frac{\left| v_{k \to i} - v_{k \to j} \right|}{N}
\end{aligned}
\tag{3.1}
$$

Let g be the multiple set of non-faulty readings, which may contain noise. We define $\delta(g) \triangleq \max(g) - \min(g)$ as the maximum reading difference between any two non-faulty PEs. If we use ε_{NA} to represent the fusion precision of naive averaging, we have

$$\varepsilon_{NA} = \max_{\forall i,j} |\tilde{v}_i - \tilde{v}_j| = \max_{\forall i,j} \frac{|v_{k \to i} - v_{k \to j}|}{N} \tag{3.2}$$

Naive averaging is not a fault tolerant algorithm, because one malicious PE can cause arbitrarily large disagreement among non-faulty PEs. Precision of naive averaging is unbounded, even if there is only one faulty PE. Therefore, there is no need to consider the general case where multiple PEs are faulty.

3.4.3 Approximate Byzantine Agreement

To reduce the impact of false inputs, Dolev et al. [5] proposed the approximate Byzantine agreement, which filters out extreme inputs.

3.4.3.1 Algorithm Introduction

Values received by PE$_i$ are sorted, giving an ordered input multiple set $V_i = \{v_{1 \to i}, \ldots, v_{N \to i}\}$. Since malicious PEs can broadcast different values to different PEs, each PE may have a different input multiset.

PE$_i$ deploys the following equation to calculate a fused output:

$$\tilde{v}_i = f_{k,\tau}(V_i) = \text{mean}(\text{select}_k(\text{reduce}^\tau(V_i))) \tag{3.3}$$

where operation reduce$^\tau(V_i)$ removes τ largest values and τ smallest values from V_i, where τ is the number of faulty PEs. By removing the τ largest and smallest values, we remove all values that are not within the range of information provided by non-faulty sensors.

The output of reduce$^\tau(V_i)$ is sampled using select$_k(\cdot)$ starting from the minimum value, where k is the sampling interval. The fusion result is the mean of the sample. Dolev's paper [5] derives the value of k for synchronous ($k = \tau$) and asynchronous ($k = 2\tau$) problems. The convergence rate also depends on k.

3.4.3.2 Precision Bound

We now find the precision bound for approximate agreement.

Theorem 3.1 (Fusion Precision of Approximate Agreement) *Consider a distributed sensor network consisting of N PEs, out of which $\tau < N/3$ are faulty ones. The fusion precision of approximate Byzantine agreement is given by Dolev et al. [5].*

$$\varepsilon_{ABA} = \max_{\forall i,j} \left| \tilde{v}_i - \tilde{v}_j \right| = \frac{\delta(g)}{\left\lfloor \frac{N-2\tau-1}{k} \right\rfloor + 1}, \quad k \geq \tau \tag{3.4}$$

where g is the set of valid measurements ($|g| = N - \tau$); $\delta(g) = \max(g) - \min(g)$ is the maximum difference between any two measurements in g.

Proof See [5].

According to (3.4), the lower bound of ε_{ABA} is obtained when $k = \tau$:

$$\min_{k} \varepsilon_{ABA} = \left. \frac{\delta(g)}{\left\lfloor \frac{N-2\tau-1}{k} \right\rfloor + 1} \right|_{k=\tau} = \frac{\delta(g)}{\left\lfloor \frac{N-2\tau-1}{\tau} \right\rfloor + 1} \tag{3.5}$$

From (3.5), we can see that the minimum fusion precision increases as τ increases. Since $\tau = \lfloor (N - 1)/3 \rfloor$ is the maximum number of faulty PEs the algorithm can allow to obtain a correct estimate, the precision bound is given by Dolev et al. [5]:

$$\min_{k} \varepsilon_{ABA} \leq \frac{\delta(g)}{\lfloor \frac{N-2\tau-1}{\tau} \rfloor + 1} |_{\tau = \lfloor (N-1)/3 \rfloor} = \frac{\delta(g)}{2} \qquad (3.6)$$

3.4.3.3 ϵ-Approximate Agreement

The agreement precision in (3.6) can be made arbitrarily small by repeating approximate averaging multiple times. An ϵ-approximate agreement metric was defined to tolerate some inconsistency. It is proved in [5] that given an arbitrarily small positive quantity ϵ, the algorithm can achieve ϵ-*approximate agreement* after multiple rounds. That is, for any non-faulty PE$_i$ and PE$_j$:

- *Agreement*: $\min_k \varepsilon_{ABA} \leq \epsilon$
- *Validity*: $\min(g)v \leq \tilde{v}_i, \tilde{v}_j \leq \max(g)$

where *Validity* means the output of non-faulty PEs is in the range indicated by initial values of the non-faulty PEs.

3.4.4   Inexact Agreement: Fast Convergence Algorithm (FCA)

3.4.4.1   Algorithm Introduction

Mahaney and Schneider proposed inexact agreement in [7]. It is assumed in their work that the measurements of non-faulty PEs have bounded distances from the correct value. Let $\widehat{v}$ denote the true value of the parameter being measured, for any non-faulty PE$_i$

$$\max_{i} |v_i - \widehat{v}| \leq \zeta \qquad (3.7)$$

where ζ is a positive constant for the desired accuracy set in advance. The number of faults tolerated by this approach is $N/3$. If more than $N/3$ faults exist, the process degrades gracefully until there are over $2N/3$ faults. Graceful degradation is defined as either (1) stopping and giving an error message, or (2) providing results within the same accuracy and precision bounds as when fewer than $N/3$ faults are found.

Based on the assumption, they proposed a Fast Convergence Algorithm (FCA). As shown in Algorithm 4.1, each PE identifies a set of τ-*acceptable* PE values from its input multiset V. A value $v \in V$ is considered *acceptable* if $\exists s, f \in \mathbb{R}$ such that,

1. $s \leq v \leq f$;
2. $f - s \leq \zeta$; and
3. $\#(V, [s, f]) \geq N - \tau$, where $\#(V, [s, f])$ is the number of elements of V that have values in interval $[s, f]$.

These rules simply codify the concepts that:

- At most τ readings can be faulty,
- Non-faulty readings must be within ζ of the correct value, and therefore
- Two non-faulty readings must be within $\varepsilon = 2\zeta$ of each other.

The precision ε is constrained to be at most 2ζ, simply because $\widehat{v} + \zeta$ and $\widehat{v} - \zeta$ are the most extreme values that can fulfill the accuracy requirement.

Let V_{accept} be the multiset of acceptable values, the algorithm replaces the unacceptable values with $e(V_{\text{accept}})$, where $e(V_{\text{accept}})$ can be any of: average, median, or midpoint of V_{accept}. FCA returns the mean of this updated multiset:

Algorithm 3.3: FCA in one round at PE_i

1: Collect values from other PEs to form a multiset V.
2: Construct the τ-acceptable set V_{accept} from V.
3: Compute $e(V_{\text{accept}})$
4: Replace values in V that are not in V_{accept} with $e(V_{\text{accept}})$.
5: $\tilde{v}_i \leftarrow$ mean (V_{accept})
6: **return** $\tilde{v}_i$;

Algorithm 3.3 calculates a reasonable approximation of FCA in one round at a given PE.

3.4.4.2   Precision Bound

FCA gives better fusion precision than approximate Byzantine agreement [5, 7].

Theorem 3.2 (Precision Bound and Accuracy Bound of FCA) *FCA algorithm is guaranteed to converge if less than $1/3$ PEs are faulty. For any two PEs PE_i and PE_j, the one-round fusion precision bound and accuracy bound [7] are*

- Precision: $\varepsilon_{FCA} = \max_{i,j} |\tilde{v}_i - \tilde{v}_j| \leq \dfrac{2\tau}{N}\delta(g)$
- Accuracy: $\zeta_{FCA} = \max_i |\tilde{v}_i - \widehat{v}| \leq \zeta + \dfrac{\tau}{N}\delta(g)$

where κ is the accuracy requirement of inputs.

Proof See [7].

If FCA is repeated multiple times for each PE, fusion precision converges to an arbitrarily small value. However, the accuracy bound might be larger than κ and cannot converge to zero.

3.4.5 Byzantine Vector Consensus (BVC)

Vaidya et al. [35] and Mendes et al. [37] consider multidimensional Byzantine agreement problems, where the local measurement at each PE is represented as a d-dimensional vector.

We note that the BVC problem is very similar to the multidimensional fusion approach presented by Brooks and Iyengar in [22]. BVC tolerates more faults $\tau = \frac{N-1}{d+2}$ than the algorithm in [22] $\tau = \frac{N}{2d}$ at the cost of performing multiple communications rounds. Brooks and Iyengar find the optimal region possible using one single communications round with complexity $O(\tau^d N \log N)$. Every round of BVC communications requires every PE to exchange data with every other PE, which has complexity $O(N^2)$.

3.4.5.1 Algorithm Introduction

Mendes used Euclidean distance between vectors to measure agreement, while Vaidya calculated distances between each element in the vector (i.e., used the $H - \infty$ metric) [35]. Both exact and approximate Byzantine Vector Consensus (BVC) algorithms are considered. We discuss Vaidya's work, since their algorithm requires a smaller number of rounds to converge [35, 37] and the underlying logic is similar.

For approximate BVC, each PE receives a d-dimensional vector valued reading from every other PEs. All PEs attempt to converge to a common d-dimensional point value v. PE_i maintains its own estimate $\tilde{v}_i$. The set of received point values is V. Approximate BVC redefines ϵ-approximation agreement for vector calculations. This agreement is met if:

- The distances between each element of any two non-faulty input vectors are no larger than the predefined constant $\epsilon > 0$.
- The vector value $\mathbf{v}_i$ being maintained at each non-faulty PE_i is in the convex hull formed by the vector values in set V from non-faulty PEs.

Let $\mathbf{V}$ be a set of vectors. We now define function $\Gamma(\mathbf{V})$

$$\Gamma(\mathbf{V}) = \cap_{\mathbf{V}' \subseteq \mathbf{V}, |\mathbf{V}'| = |\mathbf{V}| - \tau} \mathcal{H}(\mathbf{V}') \tag{3.8}$$

where $\mathcal{H}(\mathbf{V}')$ is the convex hull of $\mathbf{V}'$, and $\Gamma(\mathbf{V})$ gives a convex hull formed by the intersection of all subsets of vector measurements from $\mathcal{H}(\mathbf{V}')$ which excludes at most τ subsets.

The approximate BVC algorithm is[3]:

Algorithm 3.4: Approximate BVC in one step at one PE

1: Each PE collects values from other PEs and forms a multiset V.
2: Initialize a set of points Z to null.
3: **for** each $C \subseteq V$ such that $|C| = N - \tau$ **do**
4:   Construct $\Gamma(C)$ and choose a point deterministically[5] from $\Gamma(C)$ and add it to Z.
5: **end for**
6: **return** $\mathbf{v}' = \frac{\sum_{z \in Z} z}{|Z|}$.

Algorithm 3.4 gives a reasonable approximation of BVC in one step at a given PE.

Note that $|*|$ denotes the size of multiset $*$. V contains at least $N - \tau$ elements. $|Z| \leq C_N^{N-\tau}$.

3.4.5.2 Precision Bound

Vaidya et al. [35] proved the precision bound by finding two most divergent PEs. We show here the precision shrink of one round between $t - 1$ and t. We first introduce some notations.

- t denotes the number of rounds.
- $\mathbf{v}_i[t]$ is the vector of PE_i in tth round.
- $\mathbf{v}_{il}[t]$ is the lth element of $\mathbf{v}_i[t]$, where $1 \leq l \leq d$.
- $\Omega_l[t] = \max_{1 \leq k \leq m} \mathbf{v}_{kl}[t]$ denoted the maximum lth element of non-faulty PEs.
- $\mu_l[t] = \min_{1 \leq k \leq m} \mathbf{v}_{kl}[t]$ in m denotes the minimum lth of non-faulty PEs.

Theorem 3.3 (Precision Bound of BVC) *The maximum distance between any elements in two decision vectors after one round is*

$$\Omega_l[t] - \mu_l[t] \leq (1 - \gamma)(\Omega_l[t - 1] - \mu_l[t - 1]) \qquad (3.9)$$

where $\gamma = \dfrac{1}{N\left(\frac{N}{N-\tau}\right)}$ *and* $1 \leq \gamma \leq 1$.

Proof Please see Appendix 1.

Theorem 3.3 shows that the maximum difference any two corresponding elements in $\mathbf{v}_i (1 \leq i \leq N)$ reduce by a scale factor $(1 - \gamma)$ for each round. This

---

[3]The authors of [35] wanted their algorithm to work for any desired linear programming optimization objective function, so they left the exact function undefined. What is important is that each PE uses the same deterministic logic to choose a point from the interior of $\Gamma(C)$.

guarantees that the system will converge to a final precision of at least the predefined value ϵ.

All the other algorithms discussed in this paper only consider one round of data exchange and fusion. BVC will converge to an arbitrary predefined precision value given enough number of rounds. This is consistent with the known results for the more general Byzantine generals problem [10].

3.4.6   Marzullo's Algorithm

3.4.6.1   Algorithm Introduction

Marzullo [8] proposed an interval-based agreement sensor fusion algorithm. It is assumed in Marzullo's algorithm that each non-faulty PE gives an interval measurement that contains the correct value $\hat{v}$. We call this the validity condition of Marzullo's algorithm. Let $[l_{j \to i}, h_{j \to i}]$ represent the measurement sent to PE_i by PE_j, the collected measurements at PE_i is $V_i = \{[l_{1 \to i}, h_{1 \to i}], \ldots, [l_{N \to i}, h_{N \to i}]\}, 1 \le i \le N$. Measurements of non-faulty sensors, by definition, must overlap each other, because they all contain the correct value.

We construct a *Weighted Region Diagram (WRD)* to illustrate the scenario. An example is shown in Fig. 3.4. The following terms are used to describe a WRD:

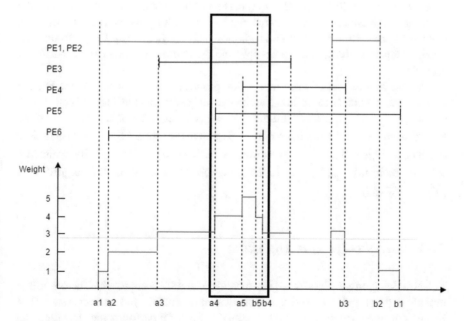

Fig. 3.4 Interval fusion process

- *Interval $I_w = [p_w, q_w]$*: A continuous area overlapped by the same w measurements.
- *Weight*: The number of overlapped measurements of an interval. Interval I_w has weight w.
- *Region $[a_w, b_w]$*: A continuous area consisting of consecutive intervals with weights no smaller than w. a_w is the left endpoint of the region and b_w is the right endpoint of the region.

Consider the WRD in Fig. 3.4, where $N = 6, \tau = 2$. Assume PE_1 and PE_2 are malicious PEs, whose measurements are represented by red bars at the top of the figure, and valid measurements are represented by black bars. The union of all measurements is divided into multiple consecutive intervals. Each interval has a height that is equal to the associated weight along the y-axis. The WRD is obtained by concatenating the intervals together. The red stair-step line outlines WRD^V constructed using the six measurements. The range that consists of consecutive intervals with weights no smaller than a given value is called a *region*. For example, region $[a_4, b_4]$ (in the bold black square frame) consists of intervals with weights no smaller than 4.

With the WRD built using the collected measurements V, each PE outputs a region $[a_{N-\tau}^V, b_{N-\tau}^V]$, which must contain the interval overlapped by $N - \tau$ nonfaulty PEs. Accordingly, $[a_{N-\tau}^V, b_{N-\tau}^V]$ must contain the correct value.

Theorem 3.4 (Fusion Precision of Marzullo's Algorithm) *For $\forall g \in G$, the precision bound of Marzullo's algorithm is region $[a_{N-2\tau}^g, b_{N-2\tau}^g]$ in WRD^g.*

Proof Since Marzullo's algorithm outputs an interval, we define its precision bound as the smallest interval that is guaranteed to contain all possible outputs, which is defined by the leftmost output and rightmost output. Since each PE performs data fusion independently, the leftmost output and the rightmost output can be calculated separately.

We first find the set of false inputs that produces the rightmost output. For the τ false measurements to be included in the output, it must at least intersect with region $[a_{N-2\tau}^g, b_{N-2\tau}^g]$. To obtain the rightmost output, it is not hard to see that the false readings should have their right ends no smaller than $b_{N-2\tau}^g$, and have their left ends no bigger than $b_{N-\tau}^g$. Accordingly, we have $[a_{N-\tau}^g, b_{N-2\tau}^g]$. By symmetry, we can obtain $[a_{N-2\tau}^g, b_{N-\tau}^g]$. So the precision bound of Marzullo's algorithm is $[a_{N-2\tau}^g, b_{N-2\tau}^g]$.

3.4.7 Brooks–Iyengar Algorithm

The Brooks–Iyengar algorithm [56] extends Marzullo's approach. The algorithm outputs a point estimate and a region that must contain the correct result. The Brooks–Iyengar algorithm is in Algorithm 4.3. Each PE performs the algorithm and yields a fused output.

The number of faults tolerated by this algorithm is the same as that provided by Marzullo's approach, $N/2$. The use of intervals in finding agreement is built on Marzullo's work. However, this algorithm can also fail gracefully, like Mahaney and Schneider, as long as fewer than $2N/3$ PEs are faulty.

Algorithm 3.5: Brooks-Iyengar Distributed Sensing Algorithm
***

Input:
    Each PE$_i$ starts with a set of measurements
    $V_i = \{[l_{1 \to i}, h_{1 \to i}], \ldots, [l_{N \to i}, h_{N \to i}]\}$ received from all PEs.
Output:
    A point estimate and a region that must contain the correct value..

1: Construct WRD^{V_i};
2: Remove regions with weight less than $N - \tau$, where τ is the number of faulty PEs.
3: The set of remaining regions is $\{[p_i^1, q_i^1], \ldots, [p_i^M, q_i^M]\}$. The weight of region $[p_i^m, q_i^m]$, $1 \le m \le M$ is denoted as w_i^m. It is the number of PE intervals that overlap with the interval.
4: Calculate point estimate $\tilde{v}_i$ of PE$_i$ as:

$$\tilde{v}_i = \frac{\sum\limits_{m=1}^{M} \frac{(p_i^m + q_i^m)}{2} \cdot w_i^m}{\sum\limits_{m=1}^{M} w_i^m} \tag{3.10}$$

and the output region that includes correct value $\hat{v}$ is
$$\left[a_{N-\tau}^{V_i}, b_{N-\tau}^{V_i} \right] = \left[\min\left(p_i^* \right), \max\left(q_i^* \right) \right]$$

Algorithm 3.5 presents the Brooks–Iyengar distributed sensing algorithm.

Now we consider an example of 5 PEs, in which PE$_5$ broadcasts different faulty measurements to other PEs. Table 3.3 gives the collected measurements at PE$_1$.

The WRD constructed using V_1 is shown in Fig. 3.5. We can write $V_1^{\tau=1}$ according to Algorithm 4.3:

$$V_1^{\tau=1} = \{([1.5, 2.7], 4), ([2.7, 2.8], 5), ([2.8, 3.2], 4)\} \tag{3.11}$$

$V_1^{\tau=1}$ consists of intervals where at least $4(= N - \tau = 5 - 1)$ measurements intersect. Using (4.11), the fused output of PE$_1$ is equal to

$$\frac{\left(4 \cdot \frac{1.5+2.7}{2} + 5 \cdot \frac{2.7+2.8}{2} + 4 \cdot \frac{2.8+3.2}{2}\right)}{13} = 2.625 \tag{3.12}$$

Table 3.3 Measurements received by PE$_1$ (V_1)

$v_{1 \to 1}$	$v_{2 \to 1}$	$v_{3 \to 1}$	$v_{4 \to 1}$	$v_{5 \to 1}$
[2.7, 6.7]	[0, 3.2]	[1.5, 4.5]	[0.8, 2.8]	[1.4, 4.6]

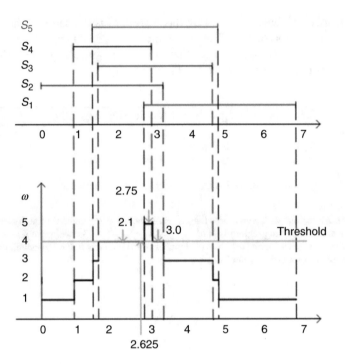

Fig. 3.5 Brooks–Iyengar algorithm in S_1

and the interval estimate is [1.5, 3.2]. Note that faulty PEs can send different measurements to non-faulty PEs and cause the fused output to disagree. We will analyze the precision bound and accuracy bound of the Brooks–Iyengar algorithm in Sect. 4.5.

3.4.7.1 Accuracy Bound

Let $\zeta_{BY} = \max_i |v_i - \hat{v}|$ be the fusion accuracy of Brooks–Iyengar algorithm. In addition to the point estimate $\tilde{v}_i$, each non-faulty PE also outputs a region $[a_{N-\tau}^{V_i}, b_{N-\tau}^{V_i}]$, which is the smallest region that is guaranteed to contain the correct value $\hat{v}$. Note that $\tilde{v}_i, \hat{v} \in [a_{N-\tau}^{V_i}, b_{N-\tau}^{V_i}]$. Hence,

$$|\tilde{v}_i - \hat{v}| \le b_{N-\tau}^{V_i} - a_{N-\tau}^{V_i}, \forall i \tag{3.13}$$

or, equivalently,

$$\zeta_{BY} = \max_i |\tilde{v}_i - \hat{v}| \le \max_i \left(b_{N-\tau}^{V_i} - a_{N-\tau}^{V_i}\right) \tag{3.14}$$

We use $\min_{\tau+1}\{|v| : v \in g\}$ to represent the length of the $(\tau + 1)$th smallest measurement for any non-faulty sensor. For example, if $g = \{[1, 14], [2, 16], [3, 18]\}$ and $\tau = 1$, then $\{|v| : v \in g\} = \{13, 14, 15\}$ and $\min_{\tau+1}\{|v| : v \in g\} = \min_{2}\{13, 14, 15\} = 14$. It is shown in [8] that

$$\max_{i} \left(b_{N-\tau}^{V_i} - a_{N-\tau}^{V_i}\right) \leq \min_{\tau+1}\{|v| : v \in g\} \tag{3.15}$$

where $\tau < \lfloor N/3 \rfloor$. From (3.14) and (3.15), we have that

$$\zeta_{BY} = \max_{i} |\tilde{v}_i - \hat{v}| \leq \min_{\tau+1}\{|v| : v \in g\} \tag{3.16}$$

3.4.7.2   Robustness

The Brooks–Iyengar algorithm may tolerate up to $\lfloor \tau/2 \rfloor$ faulty PEs. "tolerate" here means both point estimate $\tilde{v}_{BY}$ and output region are bounded by non-faulty PEs. The proof can be found in [8] (Theorem 1).

3.4.7.3   Python Code

```python
# -*- coding: utf-8 -*-
"""
Created on Mon Apr 24 15:11:52 2017

@author: aobuke
"""

import numpy as np
import matplotlib.pyplot as plt

X = np.array([[1,2],[1,2],[1.2,2.2]])

def BY(x, tau):
    N = len(x)
    xSort = np.sort(np.unique(x.ravel()))
    xMean = []
    iWeig = []
    for i in range(len(xSort)-1):
        temp = (xSort[i] + xSort[i+1])/2.0
        weight = sum((temp > x[:,0] ) & (temp < x[:,1]))
        xMean.append(temp)
        iWeig.append(weight)
        print xMean, iWeig
    xMean = np.array(xMean)
    iWeig = np.array(iWeig)
    idx = iWeig >= N - tau
```

```
27      #print xMean,iWeig,idx
28      val = sum(xMean[idx] * iWeig[idx])/sum(iWeig[idx])
29      idx2 = np.where(idx)[0]
30      print xSort, xMean, iWeig
31      return val, np.array([xSort[idx2[0]],xSort[idx2[-1]+1]])
32
33  #print BY(X, 1)
34
35  x2 = np.arange(0,10,0.2)
36  y2 = np.sin(x2)
37  N = len(x2)
38  sensor1m = y2 + np.random.random(N) - 0.5
39  sensor1Itvl = np.zeros([N,2])
40  sensor1Itvl[:,0] = sensor1m - 0.5
41  sensor1Itvl[:,1] = sensor1m + 0.5
42
43  sensor2m = y2 + np.random.random(N) - 0.5
44  sensor2Itvl = np.zeros([N,2])
45  sensor2Itvl[:,0] = sensor2m - 0.5
46  sensor2Itvl[:,1] = sensor2m + 0.5
47
48  sensor3m = y2 + np.random.random(N) - 0.5
49  sensor3Itvl = np.zeros([N,2])
50  sensor3Itvl[:,0] = sensor3m - 0.5
51  sensor3Itvl[:,1] = sensor3m + 0.5
52
53  sensor4m = y2 + 5*(np.random.random(N) - 0.5)
54  sensor4Itvl = np.zeros([N,2])
55  sensor4Itvl[:,0] = sensor4m - 0.5
56  sensor4Itvl[:,1] = sensor4m + 0.5
57
58  fusionOutput = np.zeros(N)
59  fusionBound = np.zeros([2,N])
60  for i in range(N):
61      sensorSet = np.vstack([sensor1Itvl[i,:], sensor2Itvl[i,:],
            sensor3Itvl[i,:],sensor4Itvl[i,:]])
62      print sensorSet
63      fusionOutput[i],itvl = BY(sensorSet,1)
64      fusionBound[:,i] = itvl
65
66  dBound = abs(fusionBound[0,:] - fusionOutput)
67  uBound = abs(fusionBound[1,:] - fusionOutput)
68  plt.errorbar(list(np.arange(0,50)), fusionOutput, yerr=[dBound,
        uBound])
69  plt.plot(fusionBound.T,'b-.')
70  plt.plot(y2)
71  plt.legend(['upper bound','lower bound','groundTruth','bound of
        Brooks-Iyegar Algorithm'],loc = 4)
72
73
74  plt.figure()
75  avg = (sensor4m+sensor3m+sensor2m+sensor1m)/4
76  plt.errorbar(list(np.arange(0,50)), fusionOutput, yerr=[dBound,
        uBound])
```

```
77  plt.plot(fusionBound.T,'b-.')
78  plt.plot(avg)
79  temp1 = np.where(avg > fusionBound[1,:])[0]
80  #plt.plot(temp, avg[temp], 'ro')
81  temp = np.where(avg < fusionBound[0,:])[0]
82  temp = np.hstack([temp, temp1])
83  plt.plot(temp, avg[temp], 'ro')
84  plt.legend(['upper bound','lower bound','average','outlier','
        bound of Brooks-Iyegar Algorithm'],loc = 4)
85
86
87  plt.figure()
88  temp = np.arange(y2[0],y2[0]+5,0.1)
89  tempItvl = np.zeros([len(temp),2])
90  tempItvl[:,0] = temp - 0.5
91  tempItvl[:,1] = temp + 0.5
92  avg2 = (temp+sensor3m[0]+sensor2m[0]+sensor1m[0])/4
93  fuOut = np.zeros(len(temp))
94  fuOutBound = np.zeros([2,len(temp)])
95
96  for i in range(len(temp)):
97      sensorSet = np.vstack([sensor1Itvl[0,:], sensor2Itvl[0,:],
            sensor3Itvl[0,:],tempItvl[i,:]])
98      print sensorSet
99      fuOut[i],itvl = BY(sensorSet,1)
100     fuOutBound[:,i] = itvl
101
102 dBound = abs(fuOutBound[0,:] - fuOut)
103 uBound = abs(fuOutBound[1,:] - fuOut)
104 plt.errorbar(list(temp), fuOut, yerr=[dBound,uBound])
105 plt.plot(temp,fuOutBound.T,'b-.')
106 plt.plot(temp,avg2)
107 plt.plot((temp[0],temp[-1]), (y2[0],y2[0]))
108 plt.legend(['upper bound','lower bound','average','groundTruth','
        bound of Brooks-Iyegar Algorithm'],loc=2)
109
110 #plt.plot(fusionBound.T,'b--')
111 #plt.errorbar(list(np.arange(0,50)), sensor1m, yerr=[0.5*np.ones
        (50),0.5*np.ones(50)])
112 #plt.plot(y2)
113 #plt.plot(sensor1Itvl)
```

3.5 Multidimensional Sensor Fusion

Multi-sensor fusion algorithm plays a critical role in improving the reliability of sensors. There are several motivation to fuse measurements sensed by heterogeneous sensors. These motivations include transient errors, failures caused by mechanical damages, measurement noise, and limited accuracy. Fusing the readings enables simultaneous utilization of different sensors (with different technologies) to be deployed redundantly to sense the value of a physical variable. This reduces the sensitivity of the overall networks to the failures of a single technology/sensor.

Here, we review an algorithm which obtains the best possible interpretation of partially contradictory sensor readings. Currently available algorithms return interpretations which are larger than the optimal. This has been done to avoid excessive computational complexity. The algorithm introduced here is based on data structures from computational geometry and provides the smallest possible feasible area that satisfies the constraints of the problem with a reasonable computational complexity.

Several real-world applications require accurate and time-efficient sensor measurements to enable automated data processing and make optimal decisions based on the readings of these sensors. It is worth noting that due to the multidimensional nature of the world, the collected data also includes more than one dimension. In order to reduce the sensitivity of the network to each individual sensor and increase reliability, one can increase the redundancy by installing more sensors. However, this solution increases the redundant data that is contradictory to our primary objective which is making the sensor network efficient?

Figure 3.6 shows one of these configurations. This example is the tracking system where sensor readings are a six-fold: a three dimensional vector giving the position of an object, and a three dimensional vector giving the velocity. The installed sensors scan the same area. They further ameliorate reliability using their geographical distribution. Another advantage of having multiple sensors is deploying different technologies which improves the sensing quality.

This algorithm uses the measured data by large number (N) of sensors which provide readings in an arbitrary number of dimensions (D). Further, up to a specific number (F) of the sensors may have measurements with fault. This approach utilizes range and interval trees as well as other pre-specified conditions to identify the multidimensional areas effectively. This algorithm has several applications.

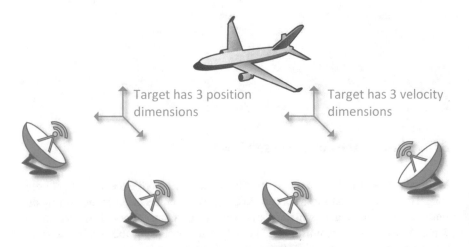

Fig. 3.6 Four independent ground stations add reliability to a target detection application

3.5.1 Faulty Sensor Averaging Problem

The introduced method was proposed by Marzullo [8]. He considers each sensor reading as an interval. Here, "concrete sensor" represents a physical sensor and its corresponding measurements. This differs from an "abstract sensor" which is the range of values readings from the "concrete sensor" are mapped to. These concepts give us a framework which can be used to deal with readings from different physical devices in a uniform manner.

Marzullo also presents the concept of a "reliable abstract sensor" which is an "abstract sensor" whose readings always contain the correct value of the physical variable within a bounded range. The "reliable abstract sensor" returns readings which are of the same shape and bounded size as readings from the "abstract sensors" and "concrete sensors." Marzullo demonstrates that a "reliable abstract sensor" can be constructed from the readings of a number of "abstract sensors" (N) if a limited number (F) of them are faulty. Multidimensional data of this kind can be presented as "d-rectangles" introduced in [9].

The systems with one dimension have been widely investigated in the literature, including Marzullo [8], and Jayasimha [50]. For instance, Chew and Marzullo have solved the multidimensional case [9] and developed a methodology that calculates the optimal area of two dimensional networks. For more than two dimensions the two existing methods, the projection method [9] and the Brooks–Iyengar method, both return regions which will often be larger than the optimal region.

3.5.2 Interval Trees

This method is designed based on a tree data structure in order to detect the intersections in D-dimensions. Six and Woods [65] provide more details of this algorithm by discussing the structure using two kinds of tree structures. They introduced range trees and interval trees. Interval trees are mainly developed to validate special conditions. Rectangles which represent sensors are placed in a binary framework to follow the areas which are limited by the maximum and minimum bounds.

3.5.3 Algorithm to Find the Optimal Region

Here, we provide the brief step-by-step algorithm for finding the optimal region proposed by Brooks et al.:

- **Step 1:** Create a primary list of cliques which are groups of intersecting measurements in the sensor network.

- **Step 2:** Re-order all of the clique points in an ascending order for the first dimension.
- **Step 3:** Move to the second dimension and repeat *Step 2*.
- **Step 4:** Two potential cases may occur at this point:
  - **Case 1:** Some of the evaluated dimensions have not faced any treats yet. For these dimensions, analyze the leaf nodes from *Step 3* and perform that step recursively.
  - **Case 2:** All dimensions faced a treat. Follow the leaves of the tree starting from the last iteration of *Step 3*.
- **Step 5:** The first dimension is effectively analyzed; *Step 3* and *Step 4* are implemented for all cliques with specific number of elements.

3.5.4 Algorithm Complexity

The complexity of *Step 1* is a constant time.

Step 2 involves sorting of $2N$ elements which takes $O(N \log N)$ time.

Creating the binary tree takes $O(N \log N)$ time.

Obtaining the regions created by the cliques can be done in $(2F + 1) * (3D - 2)$ operations which is equivalent to $O(DF)$ time.

The overall complexity depends on the recursion in steps 2–4. The equation given thus reduces to a worst-case complexity in the order of $O(N \log N * F * *D)$. Depending on design constraints, the inequality $F <= N/2D$ can be used to put this order into any two of the three essential variables.

3.5.5 Comparison with Known Methods

Table 3.4 provides a brief overview of the existing algorithms for this problem.

The Brooks–Iyengar [56] algorithm takes into account inter-dimensional effects, but may return larger than optimal regions if readings cluster significantly in all dimensions.

3.6 Conclusion

In this chapter, first, we introduced Byzantine generals problem as an example of agreement problem in Byzantine army. Then, we discussed Byzantine fault tolerance which is the resistance of a fault tolerant computer system, particularly distributed computing systems, towards electronic component failures where there is imperfect information on whether a component is failed. Afterwards, we addressed

Table 3.4 Multidimensional sensor fusion algorithm comparison

Algorithm	Dim	Complexity	Region	Comments
Marzullo	1	$O(N \log N)$	Optimal	First algorithm
Jayasimha	1	$O(N \log N)$	Optimal	List faulty readings
Chew and Marzullo if $n(S)$	2	$O(DN \log N)$	Optimal D-rectangles	Prohibitive complexity for >2 dimensions
Chew and Marzullo projection	D	$O(DN \log N)$	D-rectangles>Optimal	Ignores interdimension effects
Brooks–Iyengar	D	$O(\max(DN \log N, D**2F**2))$	D-rectangles>Optimal	Verify interdimension effects
Interval trees	D	$O(N \log N * F**D)$	Optimal	–

fault tolerant sensor fusion and how Brooks–Iyengar hybrid algorithm satisfies the requirements of both the inexact agreement problem and the sensor fusion problem. Additionally, we examined where Brooks–Iyengar algorithm stands in the literature and compare its performance with other distributed agreement algorithms theoretically. Finally, we presented a multidimensional sensor fusion algorithm proposed by Brooks and Iyengar and compare its performance with other known methods.

References

1. M.H. Amini (ed.), in *Optimization, Learning, and Control for Interdependent Complex Networks*. Advances in Intelligent Systems and Computing, vol. 2 (Springer, Cham, 2020)
2. M.H. Amini, Distributed computational methods for control and optimization of power distribution networks, PhD Dissertation, Carnegie Mellon University, 2019
3. A. Imteaj, M.H. Amini, Distributed sensing using smart end-user devices: pathway to federated learning for autonomous IoT". *Proceeding of 2019 International Conference on Computational Science and Computational Intelligence*, Las Vegas (2019)
4. M.H. Amini et al, Load management using multi-agent systems in smart distribution network, in *IEEE Power and Energy Society General Meeting* (IEEE, Piscataway, 2013), pp. 1–5
5. D. Dolev, N.A. Lynch, S.S. Pinter, E.W. Stark, W.E. Weihl, Reaching approximate agreement in the presence of faults. J. Assoc. Comput. Mach. **33**(3), 499–516 (1986)
6. L. Lamport, R. Shostak, M. Pease, The byzantine generals problem. Assoc. Comput. Mach. Trans. Program. Lang. Syst. **4**(3), 382–401 (1982)
7. S.R. Mahaney, F.B. Schneider, Inexact agreement: accuracy, precision, and graceful degradation, in *Proceedings of the fourth annual ACM symposium on Principles of distributed computing (PODC)* (1985), pp. 237–249
8. K. Marzullo, Tolerating failures of continuous-valued sensors. Assoc. Comput. Mach. Trans. Comput. Syst. **8**(4), 284–304 (1990)
9. P. Chew, K. Marzullo, Masking failures of multidimensional sensors, in *Proceedings of Tenth Symposium on Reliable Distributed Systems, 1991* (IEEE, Piscataway, 1991), pp. 32–41
10. M. Barborak, A. Dahbura, M. Malek, The consensus problem in fault-tolerant computing. Assoc. Comput. Mach. Comput. Surv. (CSUR) **25**(2), 171–220 (1993)
11. T. Krol, (n, k) concept fault tolerance. IEEE Trans. Comput. **35**(4), 339–349 (1986)

12. D. Dolev, The byzantine generals strike again. J. Algorithm. **3**(1), 14–30 (1982)
13. M.J. Fischer, N.A. Lynch, A lower bound for the time to assure interactive consistency. Inf. Process. Lett. **14**(4), 183–186 (1982)
14. R. Brooks, S.S. Iyengar, Optimal matching algorithm for multi-dimensional sensor readings, in *SPIE Proceeding Sensor Fusion and Advanced Robotics* (SPIE, Bellingham, 1995), pp. 91–99
15. A.D. Fekete, Asymptotically optimal algorithms for approximate agreement, in *Proceedings of the Fifth Annual ACM Symposium on Principles of Distributed Computing (PODC)* (1986), pp. 73–87
16. E.F. Nakamura, A.A.F. Loureiro, A.C. Frery, Information fusion for wireless sensor networks: Methods, models, and classifications. Assoc. Comput. Mach. Comput. Surv. **39**(3), 9 (2007)
17. S.S. Blackman, T.J. Broida, Multiple sensor data association and fusion in aerospace applications. J. Robot. Syst. **7**(3), 445–485 (1990)
18. R.C. Luo, M.G. Kay, Multisensor integration and fusion in intelligent systems. IEEE Trans. Syst. Man Cybern. **19**(5), 901–931 (1989)
19. F. Koushanfar, M. Potkonjak, A. Sangiovanni-Vincentell, Fault tolerance techniques for wireless ad hoc sensor networks, in *Proceedings of IEEE Sensors, 2002*, vol 2 (IEEE, Piscataway, 2002), pp. 1491–1496
20. T. Clouqueur, K.K. Saluja, P. Ramanathan, Fault tolerance in collaborative sensor networks for target detection. IEEE Trans. Comput. **53**(3), 320–333 (2004)
21. X. Luo, M. Dong, Y. Huang, On distributed fault-tolerant detection in wireless sensor networks. IEEE Trans. Comput. **55**(1), 58–70 (2006)
22. R.R. Brooks, S.S. Iyengar, *Multi-Sensor Fusion: Fundamentals and Applications with Software* (Prentice-Hall, Upper Saddle River, 1998)
23. S.S. Iyengar, *Scalable Infrastructure for Distributed Sensor Networks* (Springer, Berlin, 2006)
24. S.S. Iyengar, R.R. Brooks, *Distributed Sensor Networks: Sensor Networking and Applications* (CRC, Boca Raton, 2012)
25. M. Pease, R. Shostak, L. Lamport, Reaching agreement in the presence of faults. J. Assoc. Comput. Mach. **27**(2), 228–234 (1980)
26. M.J. Fischer, N.A. Lynch, M.S. Paterson, Impossibility of distributed consensus with one faulty process. J. Assoc. Comput. Mach. **32**(2), 374–382 (1985)
27. D. Dolev, C. Dwork, L. Stockmeyer, On the minimal synchronism needed for distributed consensus. J. Assoc. Comput. Mach. **34**(1), 77–97 (1987)
28. I. Abraham, Y. Amit, D. Dolev, Optimal resilience asynchronous approximate agreement, in *8th International Conference on Principles of Distributed Systems, OPODIS 2004* (2004), pp. 229–239
29. N.H Vaidya, L. Tseng, G. Liang, Iterative approximate byzantine consensus in arbitrary directed graphs, in *Proceedings of the 2012 ACM symposium on Principles of distributed computing* (ACM, New York, 2012), pp. 365–374
30. L. Tseng, N.H. Vaidya, Iterative approximate byzantine consensus under a generalized fault model, in *Proceedings of 14th International Conference Distributed Computing and Networking, ICDCN 2013* (2013), pp. 72–86
31. L. Su, N. Vaidya, Reaching approximate byzantine consensus with multi-hop communication, in *Proceedings of Stabilization, Safety, and Security of Distributed Systems—17th International Symposium, SSS 2015* (2015), pp. 21–35
32. B. Charron-Bost, M. Függer, T. Nowak, Approximate consensus in highly dynamic networks: the role of averaging algorithms, in *Proceedings of Part II Automata, Languages, and Programming—42nd International Colloquium, ICALP 2015* (2015), pp. 528–539
33. C. Li, M. Hurfin, Y. Wang, Brief announcement: Reaching approximate byzantine consensus in partially-connected mobile networks, in *Proceedings of Distributed Computing—26th International Symposium, DISC 2012* (2012), pp. 405–406
34. C. Li, M. Hurfin, Y. Wang, Approximate byzantine consensus in sparse, mobile ad-hoc networks. J. Parallel Distrib. Comput. **74**(9), 2860–2871 (2014)
35. N.H. Vaidya, V.K. Garg, Byzantine vector consensus in complete graphs, in *PODC* (2013), pp. 65–73

36. N.H. Vaidya, Iterative byzantine vector consensus in incomplete graphs, in *Proceedings of 15th International Conference Distributed Computing and Networking, ICDCN 2014* (2014), pp. 14–28
37. H. Mendes, M. Herlihy, Multidimensional approximate agreement in byzantine asynchronous systems, in *Proceedings of the 45th annual ACM symposium on Symposium on theory of computing(STOC)* (ACM, New York, 2013), pp. 391–400
38. A. Doudou, A. Schiper, Muteness detectors for consensus with byzantine processes, in *Proceedings of the Seventeenth Annual ACM Symposium on Principles of Distributed Computing, PODC '98* (ACM, New York, 1998)
39. N.F. Neves, M. Correia, P. Verissimo, Solving vector consensus with a wormhole. IEEE Trans. Parallel Distrib. Syst. **16**(12), 1120–1131 (2005)
40. K. Marzullo, S. Owicki, Maintaining the time in a distributed system, in *Proceedings of the second annual ACM symposium on Principles of distributed computing* (ACM, New York, 1983), pp. 295–305
41. K.A. Marzullo, *Maintaining the Time in a Distributed System: An Example of a Loosely-coupled Distributed Service (Synchronization, Fault-tolerance, Debugging)*. (PhD thesis, USA, 1984), AAI8506272
42. S.S. Iyengar, D.N. Jayasimha, D. Nadig, A versatile architecture for the distributed sensor integration problem. IEEE Trans. Comput. **43**(2), 175–185 (1994)
43. D.N. Jayasimha, S.S. Iyengar, R. Kashyap, Information integration and synchronization in distributed sensor networks. IEEE Trans. Syst. Man Cybern. **21**(5), 1032–1043 (1991)
44. D. Nadig, S.S. Iyengar, D.N. Jayasimha, A new architecture for distributed sensor integration, in *Proceedings of IEEE Southeastcon'93* (IEEE, Piscataway, 1993), p. 8
45. P. Blum, L. Meier, L. Thiele, Improved interval-based clock synchronization in sensor networks, in *Third International Symposium on Information Processing in Sensor Networks, 2004. IPSN 2004* (IEEE, Piscataway, 2004), pp. 349–358
46. L. Prasad, S.S. Iyengar, R. Kashyap, R.N. Madan, Functional characterization of fault tolerant integration in distributed sensor networks. IEEE Trans. Syst. Man Cybern. **21**(5), 1082–1087 (1991)
47. S.S. Iyengar, L. Prasad, A general computational framework for distributed sensing and fault-tolerant sensor integration. IEEE Trans. Syst. Man Cybern. **25**(4), 643–650 (1995)
48. L. Lamport, Synchronizing Time Servers (Systems Research Center, Palo Alto, 1987)
49. U. Schmid, K. Schossmaier, How to reconcile fault-tolerant interval intersection with the Lipschitz condition. Distrib. Comput. **14**(2), 101–111 (2001)
50. D.N. Jayasimha, Fault tolerance in a multisensor environment, in *Proceedings of 13th Symposium on Reliable Distributed Systems, 1994* (IEEE, Piscataway, 1994), pp. 2–11
51. D. Desovski, Y. Liu, B. Cukic, Linear randomized voting algorithm for fault tolerant sensor fusion and the corresponding reliability model, in *Ninth IEEE International Symposium on High-Assurance Systems Engineering, 2005. HASE 2005* (IEEE, Piscataway, 2005), pp. 153–162
52. B. Parhami, Distributed interval voting with node failures of various types, in *IEEE International Parallel and Distributed Processing Symposium, 2007 (IPDPS 2007)* (IEEE, Piscataway, 2007), pp. 1–7
53. L. Prasad, S.S. Iyengar, R. Rao, R.L. Kashyap, Fault-tolerant integration of abstract sensor estimates using multiresolution decomposition, in *Proceedings of IEEE Systems Man and Cybernetics Conference*, vol 5 (IEEE, Piscataway, 1993), pp. 171–176
54. L. Prasad, S.S. Iyengar, R.L. Rao, R.L. Kashyap, Fault-tolerant sensor integration using multiresolution decomposition. Phys. Rev. E **49**(4), 3452 (1994)
55. H. Qi, S.S. Iyengar, K. Chakrabarty, Distributed multi-resolution data integration using mobile agents, in *IEEE Proceedings of Aerospace Conference, 2001*, vol 3 (IEEE, Piscataway, 2001), pp. 3–1133
56. R.R. Brooks, S.S. Iyengar, Robust distributed computing and sensing algorithm. Computer **29**(6), 53–60 (1996)

57. R.R. Brooks, S.S. Iyengar, Methods of approximate agreement for multisensor fusion, in *SPIE's 1995 Symposium on OE/Aerospace Sensing and Dual Use Photonics* (International Society for Optics and Photonics, Bellingham, 1995), pp. 37–44

58. V. Kumar, Impact of Brooks–Iyengar distributed sensing algorithm on real time systems. IEEE Trans. Parallel Distrib. Syst. **25**(5), 1370–1370 (2013)

59. L. Xiao, S. Boyd, S. Lall, A scheme for robust distributed sensor fusion based on average consensus, in *Fourth International Symposium on Information Processing in Sensor Networks, 2005. IPSN 2005* (IEEE, Piscataway, 2005), pp. 63–70

60. R. Olfati-Saber, J.A. Fax, R.M. Murray, Consensus and cooperation in networked multi-agent systems. Proc. IEEE **95**(1), 215–233 (2007)

61. R. Olfati-Saber, R.M. Murray, Consensus problems in networks of agents with switching topology and time-delays. IEEE Trans. Autom. Control **49**(9), 1520–1533 (2004)

62. R.R. Murphy, Dempster-Shafer theory for sensor fusion in autonomous mobile robots. IEEE Trans. Robot. Autom. **14**(2), 197–206 (1998)

63. H. Wu, M. Siegel, R. Stiefelhagen, J. Yang, Sensor fusion using Dempster-Shafer theory [for context-aware HCI], in *Proceedings of the 19th IEEE Instrumentation and Measurement Technology Conference, 2002. IMTC/2002* (IEEE, Piscataway, 2002), vol 1, pp. 7–12

64. V.V.S. Sarma, S. Raju, Multisensor data fusion and decision support for airborne target identification. IEEE Trans. Syst. Man Cybern. **21**(5), 1224–1230 (1991)

65. H.-W. Six, D. Wood, Counting and reporting intersections of d-ranges. IEEE Trans. Comput. **31**(3), 181–187 (1982)

Part II
Advances of Sensor Fusion Algorithm

Chapter 4
Theoretical Analysis of Brooks–Iyengar Algorithm: Accuracy and Precision Bound

General Comments The theoretical feature of Brooks–Iyengar algorithm is very important for analysis and application. A critical issue in evaluating sensor fusion algorithms is finding the proper evaluation criteria. Two related, but essentially different parameters used to evaluate fusion algorithms are: *accuracy* and *precision*. In distributed sensor fusion:

- *Accuracy*, denoted by δ, measures the difference between the fusion output and the ground truth value being measured [1].
- *Precision*, denoted by ε, measures the degree of disagreement between fusion outputs of different sensors due to noise and false inputs [2].

Both concepts are important and inter-related. *Accuracy* and *precision* form the consensus paradigm of distributed systems, and are key performance indicators of a distributed system. *Accuracy* focuses on approximating the true value, which is important in applications like parameter estimation, data aggregation, sensor fusion, etc. *Precision* ensures system consistency and is essential to applications like clock synchronization, robot convergence and gathering, distributed voting, etc. Accuracy and precision are both considered critical in time-triggered architecture and cyber-physical systems safety [3, 4]. In this chapter, we establish precision bounds for several widely used distributed fault tolerant sensor fusion algorithm.

The following article with permission has been reproduced from the original copy: Ao, Buke, et al. "On precision bound of distributed fault tolerant sensor fusion algorithms." *ACM Computing Surveys (CSUR) 49.1 (2016): 5.*

© Springer Nature Switzerland AG 2020
P. Sniatala et al., *Fundamentals of Brooks–Iyengar Distributed Sensing Algorithm*,
https://doi.org/10.1007/978-3-030-33132-0_4

4.1   Introduction

Sensing applications are limited by a number of physical constraints, which makes fault tolerant sensor fusion a crucial problem. Sensors extract information about targets traversing physical media, including ambient gas, fluid, or solid media. If the medium is homogeneous and has no obstacles, stochastic models can represent the sensor readings. The statistical combination of data samples, with increasing number of inputs, usually increases the system's output accuracy and confidence while bounding its variance. In most conditions, adding additional inputs will shrink the size of the associated confidence interval.

Unfortunately, the ambient medium is rarely uniform, which introduces additional factors, such as multi-path fading, shadowing, and occlusion. A large number of noise factors exist in typical sensing scenarios and these factors are usually not consistent with the assumptions used to create the statistical models, which limits the practical utility of statistics.

In physical systems, component failures can lead to arbitrary inputs instead of small magnitude noise of uniform variance. Loss of network connectivity can remove arbitrary subsets of inputs from calculations. Other networking errors can arbitrarily modify the inputs of any sensor traversing the affected network region [5]. Sensors are generally distributed, so their readings are from different perspectives and subjected to different environmental influences.

To counteract system errors, Lamport et al. [6] proposed the Byzantine Generals Problem (BGP) where a number of decision makers strive to make the same decision in the presence of a limited number of purposely deceptive inputs. If there exists an approach that can reach correct consensus under these conditions, the system is considered robust. The BGP can be considered as a fault tolerant logic problem. Lamport et al. proved that agreement on the correct answer is always possible as long as (ignoring network topology constraints) less than $1/3$ of inputs are faulty. The original problem considered a binary choice. Later researchers considered problems that include continuous variables.

Fault tolerant sensor fusion has the same goal as BGP, which is to achieve agreement (consensus) among the processing elements (PEs) in the presence of faulty, noisy, or malicious data. In this paper, PE is used to indicate a smart, networked sensing device capable of sensing, receiving inputs, fusing inputs, and broadcasting its results.

A critical issue in evaluating sensor fusion algorithms is finding the proper evaluation criteria. Two related, but essentially different parameters used to evaluate fusion algorithms are: *accuracy* and *precision*.[1] In distributed sensor fusion:

---

[1]Note that the literature has not always been consistent in using these terms. The reader is warned to not assume that our use of these terms and associated variable notations will match those in the papers we reference.

- *Accuracy*, denoted by δ, measures the difference between the fusion output and the ground truth value being measured [1].
- *Precision*, denoted by ε, measures the degree of disagreement between fusion outputs of different sensors due to noise and false inputs [2].

Both concepts are important and inter-related. *Accuracy* and *precision* form the consensus paradigm of distributed systems, and are key performance indicators of a distributed system. *Accuracy* focuses on approximating the true value, which is important in applications like parameter estimation, data aggregation, sensor fusion, etc. *Precision* ensures system consistency and is essential to applications like clock synchronization, robot convergence and gathering, distributed voting, etc. Accuracy and precision are both considered critical in time-triggered architecture and cyber-physical systems safety [3, 4].

Compared with *accuracy* that has been widely used to evaluate the performance of fusion algorithms, *precision*, especially *precision bounds*, which express the worst-case (largest possible) disagreement among PE outputs, is much less studied. In this paper, we establish precision bounds for several widely used distributed fault tolerant sensor fusion algorithm. The objective is to help users select the appropriate fusion approach, fitting their input data types and output requirements.

The rest of the paper is organized as follows. Section 4.2 discusses related work and briefly introduces the algorithms investigated in this paper. Section 4.3 looks at the precision of these algorithms. Three selected algorithms are compared in Sect. 4.4. Section 4.5 gives formal proof of the precision bound of the Brooks–Iyengar algorithm. The paper concludes in Sect. 4.6 with a summary of the investigated algorithms.

4.2 Background

Consensus in distributed systems and sensor fusion are fundamental problems. Applications of these problems exist in time synchronization, sensor networks, aerospace, defense, and other domains [7–10]. Fault tolerance is a critical issue [11–13] and myriad algorithms have been proposed for distributed sensing, information fusion, and time synchronization [5, 7, 9, 12, 14, 15].

To formally express this problem, let us consider a network of N PEs, $\tau < N$ of the PEs may be faulty and provide erroneous data. To solve this problem, we assume that the faulty PEs can maliciously conspire to create the worst possible set of inputs to force the algorithm to fail. If the approach can function properly, even in response to this set of inputs, then it effectively solves the problem. Concentrating on worst-case analysis also allows us to establish effective performance bounds.

Consider a set of distributed PEs, each PE_i ($i \leq N$) measures some parameter with a noise factor that introduces a random, bounded deviation from the correct value. An unknown subset of the PEs are faulty. A faulty PE can broadcast an arbitrary measurement to each collaborating PE. Ignoring the network topology, for the system of N PEs to be able to reach consensus, the number of false inputs must be less than $N/3$. A proof of this limit is in [6].

The PEs exchange values, the effects of the network topology on fusion are outside the scope of this paper. Let $\mathscr{V} = \{v_{1 \to i}, v_{2 \to i}, \ldots, v_{N \to i}\}$ be the set of sensor readings received by PE$_i$. PE$_i$'s input from PE$_j$ is $v_{j \to i}$. In some scenarios, inputs are scalar values; in other cases, inputs are intervals expressed as $v_{j \to i} = [l_{j \to i}, h_{j \to i}]$, where $l_{i \to i}$ represents the lower bound and $h_{j \to i}$ represents the higher bound. If all PEs are non-faulty, any deterministic algorithm would reach consensus, since each PE runs the same algorithm with the same set of inputs.

Values sent from faulty PEs can arbitrarily differ from PE to PE. To determine whether or not the results at each PE$_i$ will converge, it is useful to consider directly the worst-case scenario where faulty PEs transmit values in a way that maximizes the difference between PE results. The agreement algorithm must minimize the influence of the false inputs, in order to reach consensus and maximize the precision of the algorithm results. Fusion algorithm agreement precision is defined as:

Definition 4.1 (Agreement Precision) Agreement precision of fusion algorithms for different types of inputs is[2]

$$
\begin{cases}
\max\limits_{\forall i,j=1,\cdots N} \left\{ \left| \tilde{v}_i - \tilde{v}_j \right| \right\}, \text{ for scalar inputs,} \\
\max\limits_{\forall i,j=1,\cdots N} \left\{ \left| \tilde{v}_i^d - \tilde{v}_j^d \right| \right\}, \text{ for the } d\text{th element of vector inputs,} \\
\max\limits_{\forall i,j=1,\cdots N} \left\{ \left| \tilde{v}_i^h - \tilde{v}_j^l \right|, \left| \tilde{v}_j^h - \tilde{v}_i^l \right| \right\}, \text{ for interval inputs}
\end{cases}
\tag{4.1}
$$

where $\tilde{v}_i$ is the scalar fused output of PE$_i$ and $\tilde{v}_j$ is the scalar fused output of PE$_j$; $\tilde{v}_i^l/\tilde{v}_i^h$ is the lower/upper bound of the interval-valued fused output of PE$_i$ and $\tilde{v}_j^l/\tilde{v}_j^h$ is the lower/upper bound of the interval-valued fused output of PE$_j$.

We now review the set of fusion approaches most relevant to this problem:

- *The Byzantine Generals Problem (BGP)* for synchronous systems was addressed in [16] and [6]. However, Fischer [17] proved that, in asynchronous systems, it is in general not possible to guarantee convergence. In [2], the authors showed that convergence was possible in partially asynchronous systems, with the presence of some synchrony. *Dolev et al.'s approximate agreement* algorithm reached consensus within known precision bounds. Fekete [18] modified Dolev's approach to have faster convergence.
- *Mahaney and Schneider's inexact agreement* considered both accuracy and precision. Abraham [19] presented a deterministic optimal resilience approximate agreement algorithm that can tolerate more Byzantine faults than Dolev's work [20]. More recently, Vaidya studied the iterative approximate Byzantine

---

[2]RRB can remove $i \neq j$, since if $i == j$ you get zero for scalar. For interval, it does not matter whether or not they are the same thing. For interval, you only need one (upper bound–lower bound). We do not specify the order we look at i and j.

consensus (IABC) algorithm to reach consensus for an arbitrary directed graph [21–23]. Approximate agreement has also been studied in dynamic networks [24–26].

- *Byzantine Vector Consensus (BVC)* reaches consensus when inputs are vectors under complete graphs [27] and incomplete graphs [28]. Mendes et al. [29] studied the multidimensional approximate agreement in asynchronous systems, where the *agreement* definition is different: the distance used in multidimensional approximate agreement is Euclidean distance. Another form of consensus in Byzantine asynchronous system is to agree on a vector, while the PEs' input is scalar [30, 31].

- *Marzullo* [32] proposed a fault tolerance fusion approach that uses interval inputs, which finds an interval where all valid readings intersect. In most cases, Marzullo's approach achieved better accuracy than individual sensor inputs. The fused interval is at least as accurate as the range of the least accurate individual sensor. Marzullo's algorithm has been used for clock synchronization [33, 34] and information integration [35–37]. Blum [38] found the worst-case for Marzullo's method in clock synchronization. To reduce the output interval width of Marzullo's method, Prasad's method chose the interval where most input intervals intersect [39, 40]. In order to satisfy the Lipschitz conditions that minor changes of input cause only minor changes of output [41], Schmid proposed [42] that the output interval might be slightly larger than Marzullo's method. Jayasimha [43] extended the original Marzullo's method to detect sensors and combinations of sensors that may be faulty. Chew and Marzullo extended the original Marzullo's method from one dimension to multi-dimension [44]. Besides Marzullo's method and its related works, Desovski [45] proposed a randomized voting algorithm to choose the interval that has votes larger than v, where v is determined by numbers of faulty sensors. Parhami [46] considered interval voting that combines either preference or uncertainty intervals.

- *Multi-Resolution Decomposition* techniques have been applied to fault tolerance. Prasad [47, 48] employed multi-resolution integration (MRI) to recognize and isolate the most prominent and robust peaks in a region for fault-tolerant sensor integration when the numbers of sensors are large and large portion of sensor faults are tame. Qi [49] extended the original MRI for Mobile-Agent-based distributed sensor network.

- *The Brooks–Iyengar algorithm* [50–52] was used as a distributed tracking algorithm in the DARPA SenseIT program and was then applied to a real-time extension of Linux [53]. This approach considers intervals where $N - \tau$ intervals overlap and performs a weighted average of the interval midpoints. This minimizes the influence of faulty inputs by only considering ranges where faulty inputs agree with a number of valid inputs.

Instead of assuming the presence of malicious inputs, other fusion approaches assume data is contaminated by a limited amount of noise [54–56] that is Gaussian. Fusion typically uses tools from probability, such as maximum likelihood estimation. The PEs try to agree on a value only by iteratively exchanging information with neighbors. Sensor fusion by Dempster–Shafer Theory in areas of autonomous mobile robots, context sensing and target identification has also been proposed [57–59].

4.3 Precision Bounds

In this article, the objective of distributed sensor fusion is to achieve consensus among the PEs and minimize the impact of faulty data. This section establishes precision bounds for five fault tolerant fusion algorithms that are representative of larger classes. The notation used in this section is in Table 4.1.

4.3.1 Naive Averaging

We start by considering one of the simplest fusion approaches. Consider a network of N PEs. Each PE broadcasts its local measurement to all the other PEs and calculates its output as the average value of the measurements it receives. Let $v_{j \to i}$ denote the measurement received by PE_i from PE_j, the output $\tilde{v}_i$ using naive averaging is

$$\tilde{v}_i = \frac{v_{1 \to i} + v_{2 \to i} + \cdots + v_{N \to i}}{N}$$

Table 4.1 List of notation

Notation	Description
$\widehat{v}$	The ground truth value being measured
N	Total number of PEs
τ	The number of faulty PEs
$v_{j \to i}$	Measurement received by PE_i from PE_j
V_i	The multiset of measurements received by PE_i, i.e., $\{v_{1 \to i}, \cdots, v_{N \to i}\}$
g	The multiset of measurements from non-faulty/valid PEs
G	The set of all possible valid measurements, so $g \in G$
$\tilde{v}_i$	Fused output at PE_i
$\delta(g)$	The maximum difference between any two non-faulty measurements, i.e., $\max(g) - \min(g)$
ε	Precision of fusion results
ζ	Accuracy of fusion results
$[l_{j \to i}, h_{j \to i}]$	The interval measurement sent to PE_i by PE_j

If we assume PE_k sends different readings to other PEs, the difference between the outputs at PE_i and PE_j is

$$
\begin{aligned}
&\left| \tilde{v}_i - \tilde{v}_j \right| \\
&= \left| \frac{v_{1\to i} + \cdots + v_{k\to i} + \cdots + v_{N\to i}}{N} - \frac{v_{1\to j} + \cdots + v_{k\to j} + \cdots + v_{N\to j}}{N} \right| \\
&= \frac{\left| v_{k\to i} - v_{k\to j} \right|}{N}
\end{aligned} \tag{4.2}
$$

Let g be the multiset of non-faulty readings, which may contain noise. We define $\delta(g) \triangleq \max(g) - \min(g)$ as the maximum reading difference between any two non-faulty PEs. If we use ε_{NA} to represent the fusion precision of naive averaging, we have

$$
\varepsilon_{NA} = \max_{\forall i,j} \left| \tilde{v}_i - \tilde{v}_j \right| = \max_{\forall i,j} \frac{\left| v_{k\to i} - v_{k\to j} \right|}{N} \tag{4.3}
$$

Naive averaging is not a fault tolerant algorithm, because one malicious PE can cause arbitrarily large disagreement among non-faulty PEs. Precision of naive averaging is unbounded, even if there is only one faulty PE. Therefore, there is no need to consider the general case where multiple PEs are faulty.

4.3.2 Approximate Byzantine Agreement

To reduce the impact of false inputs, Dolev et al. [20] proposed the approximate Byzantine agreement, which filters out extreme inputs.

4.3.2.1 Algorithm Introduction

Values received by PE_i are sorted, giving an ordered input multiset $V_i = \{v_{1\to i}, \cdots, v_{N\to i}\}$. Since malicious PEs can broadcast different values to different PEs, each PE may have a different input multiset.

PE$_i$ uses the following equation to calculate a fused output

$$
\tilde{v}_i = f_{k,\tau}(V_i) = \text{mean}(\text{select}_k(\text{reduce}^\tau(V_i))) \tag{4.4}
$$

where operation $\text{reduce}^\tau(V_i)$ removes τ largest values and τ smallest values from V_i, where τ is the number of faulty PEs. By removing the τ largest and smallest values, we remove all values that are not within the range of information provided by non-faulty sensors.

The output of reduce$^\tau (V_i)$ is sampled using select$_k (\cdot)$ starting from the minimum value, where k is the sampling interval. The fusion result is the mean of the sample. Dolev's paper [20] derives the value of k for synchronous ($k = \tau$) and asynchronous ($k = 2\tau$) problems. The convergence rate also depends on k.

4.3.2.2 Precision Bound

We now find the precision bound for approximate agreement.

Theorem 4.1 (Fusion Precision of Approximate Agreement) *Consider a distributed sensor network consisting of N PEs, out of which $\tau < N/3$ are faulty ones. The fusion precision of approximate Byzantine agreement is given by Dolev [20].*

$$\varepsilon_{ABA} = \max_{\forall i,j} \left| \tilde{v}_i - \tilde{v}_j \right| = \frac{\delta(g)}{\lfloor \frac{N-2\tau-1}{k} \rfloor + 1}, k \geq \tau \tag{4.5}$$

where g is the set of valid measurements ($|g| = N - \tau$); $\delta(g) = \max(g) - \min(g)$ is the maximum difference between any two measurements in g.

Proof See [20].

According to (4.5), the lower bound of ε_{ABA} is obtained when $k = \tau$:

$$\min_k \varepsilon_{ABA} = \frac{\delta(g)}{\lfloor \frac{N-2\tau-1}{k} \rfloor + 1} \bigg|_{k=\tau} = \frac{\delta(g)}{\lfloor \frac{N-2\tau-1}{\tau} \rfloor + 1} \tag{4.6}$$

From (4.6), we can see that the minimum fusion precision increases as τ increases. Since $\tau = \lfloor (N - 1)/3 \rfloor$ is the maximum number of faulty PEs the algorithm can allow to obtain a correct estimate, the precision bound is given by [20]

$$\min_k \varepsilon_{ABA} \leq \frac{\delta(g)}{\lfloor \frac{N-2\tau-1}{\tau} \rfloor + 1} \bigg|_{\tau=\lfloor (N-1)/3 \rfloor} = \frac{\delta(g)}{2} \tag{4.7}$$

4.3.2.3 ϵ-Approximate Agreement

The agreement precision in (4.7) can be made arbitrarily small by repeating approximate averaging multiple times. An ϵ-approximate agreement metric was defined to tolerate some inconsistency. It is proved in [20] that given an arbitrarily small positive quantity ϵ, the algorithm can achieve ϵ-*approximate agreement* after multiple rounds. That is, for any non-faulty PE$_i$ and PE$_j$:

- *Agreement*: $\min_k \varepsilon_{ABA} \leq \epsilon$.
- *Validity*: $\min(g)v \leq \tilde{v}_i, \tilde{v}_j \leq \max(g)$

where *Validity* means the output of non-faulty PEs is in the range indicated by initial values of the non-faulty PEs.

4.3.3 Inexact Agreement: Fast Convergence Algorithm (FCA)

4.3.3.1 Algorithm Introduction

Mahaney and Schneider proposed inexact agreement in [1]. It is assumed in their work that the measurements of non-faulty PEs have bounded distances from the correct value. Let $\widehat{v}$ denote the true value of the parameter being measured, for any non-faulty PE_i

$$\max_{i} |v_i - \widehat{v}| \leq \zeta, \tag{4.8}$$

where ζ is a positive constant for the desired accuracy set in advance. The number of faults tolerated by this approach is $N/3$. If more than $N/3$ faults exist, the process degrades gracefully until there are over $2N/3$ faults. Graceful degradation is defined as either (1) stopping and giving an error message, or (2) providing results within the same accuracy and precision bounds as when fewer than $N/3$ faults are found.

Based on the assumption, they proposed a fast convergence algorithm (FCA). As shown in Algorithm 4.1, each PE identifies a set of τ-*acceptable* PE values from its input multiset V. A value $v \in V$ is considered *acceptable* if $\exists s, f \in \mathbb{R}$ such that,

1. $s \leq v \leq f$;
2. $f - s \leq \zeta$; and
3. $\#(V, [s, f]) \geq N - \tau$, where $\#(V, [s, f])$ is the number of elements of V that have values in interval $[s, f]$.

These rules simply codify the concepts that:

- At most τ readings can be faulty,
- Non-faulty readings must be within ζ of the correct value, and therefore
- Two non-faulty readings must be within $\varepsilon = 2\zeta$ of each other.

The precision ε is constrained to be at most 2ζ, simply because $\widehat{v} + \zeta$ and $\widehat{v} - \zeta$ are the most extreme values that can fulfill the accuracy requirement.

Let V_{accept} be the multiset of acceptable values, the algorithm replaces the unacceptable values with $e(V_{accept})$, where $e(V_{accept})$ can be any of: average, median, or midpoint of V_{accept}. FCA returns the mean of this updated multiset:

4.3.3.2 Precision Bound

FCA gives better fusion precision than approximate Byzantine agreement [1, 20].

Algorithm 4.1: FCA in one round at PE_i

1: Collect values from other PEs to form a multiset V.
2: Construct the τ-acceptable set V_{accept} from V.
3: Compute $e(V_{\text{accept}})$
4: Replace values in V that are not in V_{accept} with $e(V_{\text{accept}})$.
5: $\tilde{v}_i \leftarrow \text{mean}(V_{\text{accept}})$
6: **return** $\tilde{v}_i$;

Theorem 4.2 (Precision Bound and Accuracy Bound of FCA) *FCA algorithm is guaranteed to converge if less than $1/3$ PEs are faulty. For any two PEs, PE_i and PE_j, the one-round fusion precision bound and accuracy bound [1] are*

- Precision: $\varepsilon_{FCA} = \max_{i,j} |\tilde{v}_i - \tilde{v}_j| \leq \dfrac{2\tau}{N}\delta(g)$
- Accuracy: $\zeta_{FCA} = \max_i |\tilde{v}_i - \widehat{v}| \leq \zeta + \dfrac{\tau}{N}\delta(g)$

where κ is the accuracy requirement of inputs.

Proof See [1].

If FCA is repeated multiple times for each PE, fusion precision converges to an arbitrarily small value. However, the accuracy bound might be larger than κ and cannot converge to zero.

4.3.4 Byzantine Vector Consensus (BVC)

Vaidya et al. [27] and Mendes et al. [29] consider multidimensional Byzantine agreement problems, where the local measurement at each PE is represented as a d-dimensional vector.

We note that the BVC problem is very similar to the multidimensional fusion approach presented by Brooks and Iyengar in [5]. BVC tolerates more faults $\tau = \dfrac{N-1}{d+2}$ than the algorithm in [5] $\tau = \dfrac{N}{2d}$ at the cost of performing multiple communications rounds. Brooks and Iyengar find the optimal region possible using one single communications round with complexity $O(\tau^d N \log N)$. Every round of BVC communications requires every PE to exchange data with every other PE, which has complexity $O(N^2)$.

4.3.4.1 Algorithm Introduction

Mendes used Euclidean distance between vectors to measure agreement, while Vaidya calculated distances between each element in the vector (i.e., used the $H - \infty$ metric) [27]. Both exact and approximate Byzantine Vector Consensus

(BVC) algorithms are considered. We discuss Vaidya's work, since their algorithm requires a smaller number of rounds to converge [27, 29] and the underlying logic is similar.

For approximate BVC, each PE receives a d-dimensional vector valued reading from every other PEs. All PEs attempt to converge to a common d-dimensional point value v. PE_i maintains its own estimate $\tilde{\mathbf{v}}_i$. The set of received point values is V. Approximate BVC redefines ϵ-approximation agreement for vector calculations. This agreement is met if:

- The distances between each element of any two non-faulty input vectors are no larger than the predefined constant $\epsilon > 0$.
- The vector value $\mathbf{v}_i$ being maintained at each non-faulty PE_i is in the convex hull formed by the vector values in set V from non-faulty PEs.

Let $\mathbf{V}$ be a set of vectors. We now define function $\Gamma(\mathbf{V})$

$$\Gamma(\mathbf{V}) = \cap_{\mathbf{V}' \subseteq \mathbf{V}, |\mathbf{V}'| = |\mathbf{V}| - \tau} \mathscr{H}(\mathbf{V}') \tag{4.9}$$

where $\mathscr{H}(\mathbf{V}')$ is the convex hull of $\mathbf{V}'$, and $\Gamma(\mathbf{V})$ gives a convex hull formed by the intersection of all subsets of vector measurements from $\mathscr{H}(\mathbf{V}')$ which excludes at most τ subsets.

The approximate BVC algorithm is:

Algorithm 4.2: Approximate BVC in one step at one PE

1: Each PE collects values from other PEs and forms a multiset V.
2: Initialize a set of points Z to null.
3: **for** each $C \subseteq V$ such that $|C| = N - \tau$ **do**
4: Construct $\Gamma(C)$ and choose a point deterministically[3] from $\Gamma(C)$ and add it to Z.
5: **end for**
6: **return** $\mathbf{v}' = \frac{\sum_{z \in Z} \mathbf{z}}{|Z|}$.

Algorithm 4.2 gives a reasonable approximation of BVC in one step at one PE. Note that $|*|$ denotes the size of multiset $*$. V contains at least $N - \tau$ elements. $|Z| \leq C_N^{N-\tau}$.

4.3.4.2   Precision Bound

Vaidya et al. [27] proved the precision bound by finding two most divergent PEs. We show here the precision shrink of one round between $t - 1$ and t. We first introduce some notations.

---

[3]The authors of [27] wanted their algorithm to work for any desired linear programming optimization objective function, so they left the exact function undefined. What is important is that each PE uses the same deterministic logic to choose a point from the interior of $\Gamma(C)$.

- t denotes the number of rounds.
- $\mathbf{v}_i[t]$ is the vector of PE_i in tth round.
- $\mathbf{v}_{il}[t]$ is the lth element of $\mathbf{v}_i[t]$, where $1 \le l \le d$.
- $\Omega_l[t] = \max_{1 \le k \le m} \mathbf{v}_{kl}[t]$ denoted the maximum lth element of non-faulty PEs.
- $\mu_l[t] = \min_{1 \le k \le m} \mathbf{v}_{kl}[t]$ in m denotes the minimum lth of non-faulty PEs.

Theorem 4.3 (Precision Bound of BVC) *The maximum distance between any elements in two decision vectors after one round is*

$$\Omega_l[t] - \mu_l[t] \le (1 - \gamma)(\Omega_l[t-1] - \mu_l[t-1]) \tag{4.10}$$

where $\gamma = \dfrac{1}{N\left(\frac{N}{N-\tau}\right)}$ *and* $1 \le \gamma \le 1$.

Proof Please see Appendix.

Theorem 4.3 shows that the maximum difference any two corresponding elements in $\mathbf{v}_i(1 \le i \le N)$ reduce by a scale factor $(1 - \gamma)$ for each round. This guarantees that the system will converge to a final precision of at least the predefined value ϵ.

All the other algorithms discussed in this paper only consider one round of data exchange and fusion. BVC will converge to an arbitrary predefined precision value given enough number of rounds. This is consistent with the known results for the more general Byzantine Generals Problem [9].

4.3.5 Marzullo's Algorithm

4.3.5.1 Algorithm Introduction

Marzullo [32] proposed an interval-based agreement sensor fusion algorithm. It is assumed in Marzullo's algorithm that each non-faulty PE gives an interval measurement that contains the correct value $\hat{v}$. We call this the validity condition of Marzullo's algorithm. Let $[l_{j \to i}, h_{j \to i}]$ represent the measurement sent to PE_i by PE_j, the collected measurements at PE_i is $V_i = \{[l_{1 \to i}, h_{1 \to i}], \cdots, [l_{N \to i}, h_{N \to i}]\}, 1 \le i \le N$. Measurements of non-faulty sensors, by definition, must overlap each other, because they all contain the correct value.

We construct a *weighted region diagram (WRD)* to illustrate the scenario. An example is shown in Fig. 4.1. The following terms are used to describe a WRD.

- *Interval $I_w = [p_w, q_w]$*: A continuous area overlapped by the same w measurements.
- *Weight*: The number of overlapped measurements of an interval. Interval I_w has weight w.
- *Region $[a_w, b_w]$*: A continuous area consisting of consecutive intervals with weights no smaller than w. a_w is the left endpoint of the region and b_w is the right endpoint of the region.

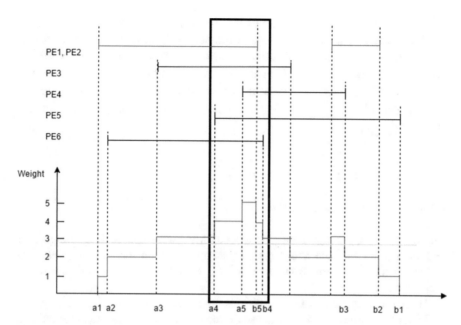

Fig. 4.1 Interval fusion process

Consider the WRD in Fig. 4.1, where $N = 6, \tau = 2$. Assume PE$_1$ and PE$_2$ are malicious PEs, whose measurements are represented by red bars at the top of the figure, and valid measurements are represented by black bars. The union of all measurements is divided into multiple consecutive intervals. Each interval has a height that is equal to the associated weight along the y-axis. The WRD is obtained by concatenating the intervals together. The red stair-step line outlines WRD^V constructed using the six measurements. The range that consists of consecutive intervals with weights no smaller than a given value is called a *region*. For example, region $[a_4, b_4]$ (in the bold black square frame) consists of intervals with weights no smaller than 4.

With the WRD built using the collected measurements V, each PE outputs a region $[a^V_{N-\tau}, b^V_{N-\tau}]$, which must contain the interval overlapped by $N - \tau$ non-faulty PEs. Accordingly, $[a^V_{N-\tau}, b^V_{N-\tau}]$ must contain the correct value.

Theorem 4.4 (Fusion Precision of Marzullo's Algorithm) *For $\forall g \in G$, the precision bound of Marzullo's algorithm is region $[a^g_{N-2\tau}, b^g_{N-2\tau}]$ in WRD^g.*

Proof Since Marzullo's algorithm outputs an interval, we define its precision bound as the smallest interval that is guaranteed to contain all possible outputs, which is defined by the leftmost output and rightmost output. Since each PE performs data fusion independently, the leftmost output and the rightmost output can be calculated separately.

We first find the set of false inputs that produces the rightmost output. For the τ false measurements to be included in the output, it must at least intersect with region $[a^g_{N-2\tau}, b^g_{N-2\tau}]$. To obtain the rightmost output, it is not hard to see that the false readings should have their right ends no smaller than $b^g_{N-2\tau}$, and have their left ends no bigger than $b^g_{N-\tau}$. Accordingly, we have $[a^g_{N-\tau}, b^g_{N-2\tau}]$. By symmetry, we can obtain $[a^g_{N-2\tau}, b^g_{N-\tau}]$. So the precision bound of Marzullo's algorithm is $[a^g_{N-2\tau}, b^g_{N-2\tau}]$.

4.3.6 Brooks–Iyengar Algorithm

4.3.6.1 Algorithm Introduction

The Brooks–Iyengar algorithm [52] extends Marzullo's approach. The algorithm outputs a point estimate and a region that must contain the correct result. The Brooks–Iyengar algorithm is in Algorithm 4.3. Each PE performs the algorithm and yields a fused output.

The number of faults tolerated by this algorithm is the same as that provided by Marzullo's approach, $N/2$. The use of intervals in finding agreement is built on Marzullo's work. However, this algorithm can also fail gracefully, like Mahaney and Schneider, as long as fewer than $2N/3$ PEs are faulty.

Now we consider an example of 5 PEs, in which PE_5 broadcasts different faulty measurements to other PEs. Table 4.2 gives the collected measurements at PE_1.

The WRD constructed using V_1 is shown in Fig. 4.2. We can write $V_1^{\tau=1}$ according to Algorithm 4.3:

Algorithm 4.3: Brooks–Iyengar distributed sensing algorithm

Input:
    Each PE_i starts with a set of measurements $V_i = \{[l_{1\to i}, h_{1\to i}], \cdots, [l_{N\to i}, h_{N\to i}]\}$ received from all PEs.

Output:
    A point estimate and a region that must contain the correct value..
    1: Construct WRD^{V_i};
    2: Remove regions with weight less than $N - \tau$, where τ is the number of faulty PEs.
    3: The set of remaining regions is $\{[p^1_i, q^1_i], \cdots, [p^M_i, q^M_i]\}$. The weight of region $[p^m_i, q^m_i], 1 \le m \le M$ is denoted as w^m_i. It is the number of PE intervals that overlap with the interval.
    4: Calculate point estimate $\tilde{v}_i$ of PE_i as:

$$\tilde{v}_i = \frac{\sum_{m=1}^{M} \frac{(p^m_i + q^m_i) \cdot w^m_i}{2}}{\sum_{m=1}^{M} w^m_i} \tag{4.11}$$

    and the output region that includes correct value $\hat{v}$ is
    $[a^{V_i}_{N-\tau}, b^{V_i}_{N-\tau}] = [\min(p^*_i), \max(q^*_i)]$

Table 4.2 Measurements received by PE$_1$ (V_1)

$v_{1 \to 1}$	$v_{2 \to 1}$	$v_{3 \to 1}$	$v_{4 \to 1}$	$v_{5 \to 1}$
[2.7, 6.7]	[0, 3.2]	[1.5, 4.5]	[0.8, 2.8]	[1.4, 4.6]

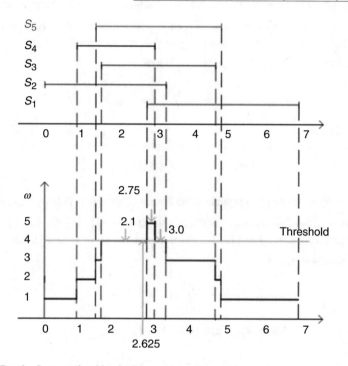

Fig. 4.2 Brooks–Iyengar algorithm in S_1

$$V_1^{\tau=1} = \{([1.5, 2.7], 4), ([2.7, 2.8], 5), ([2.8, 3.2], 4)\} \tag{4.12}$$

$V_1^{\tau=1}$ consists of intervals where at least $4 (= N - \tau = 5 - 1)$ measurements intersect. Using (4.11), the fused output of PE$_1$ is equal to

$$\frac{\left(4 \cdot \frac{1.5+2.7}{2} + 5 \cdot \frac{2.7+2.8}{2} + 4 \cdot \frac{2.8+3.2}{2}\right)}{13} = 2.625 \tag{4.13}$$

and the interval estimate is [1.5, 3.2]. Note that faulty PEs can send different measurements to non-faulty PEs and cause the fused output to disagree. We will analyze the precision bound and accuracy bound of the Brooks–Iyengar algorithm in Sect. 4.5.

4.3.6.2   Accuracy Bound

Let $\zeta_{BY} = \max_i |v_i - \hat{v}|$ be the fusion accuracy of Brooks–Iyengar algorithm. In addition to the point estimate $\tilde{v}_i$, each non-faulty PE also outputs a region $[a_{N-\tau}^{V_i}, b_{N-\tau}^{V_i}]$, which is the smallest region that is guaranteed to contain the correct value $\hat{v}$. Note that $\tilde{v}_i, \hat{v} \in [a_{N-\tau}^{V_i}, b_{N-\tau}^{V_i}]$. Hence,

$$|\tilde{v}_i - \hat{v}| \le b_{N-\tau}^{V_i} - a_{N-\tau}^{V_i}, \forall i \tag{4.14}$$

or, equivalently,

$$\zeta_{BY} = \max_i |\tilde{v}_i - \hat{v}| \le \max_i (b_{N-\tau}^{V_i} - a_{N-\tau}^{V_i}) \tag{4.15}$$

We use $\min_{\tau+1}\{|v| : v \in g\}$ to represent the length of the $(\tau + 1)$th smallest measurement for any non-faulty sensor. For example, if $g = \{[1, 14], [2, 16], [3, 18]\}$ and $\tau = 1$, then $\{|v| : v \in g\} = \{13, 14, 15\}$ and $\min_{\tau+1}\{|v| : v \in g\} = \min_2\{13, 14, 15\} = 14$. It is shown in [32] that

$$\max_i (b_{N-\tau}^{V_i} - a_{N-\tau}^{V_i}) \le \min_{\tau+1}\{|v| : v \in g\} \tag{4.16}$$

where $\tau < \lfloor N/3 \rfloor$. From (4.15) and (4.16), we have that

$$\zeta_{BY} = \max_i |\tilde{v}_i - \hat{v}| \le \min_{\tau+1}\{|v| : v \in g\} \tag{4.17}$$

4.3.6.3   Robustness

The Brooks–Iyengar algorithm may tolerate up to $\lfloor \tau/2 \rfloor$ faulty PEs. "tolerate" here means both point estimate $\tilde{v}_{BY}$ and output region are bounded by non-faulty PEs. The proof can be found in [32, Theorem 1].

4.4   Comparison

In this section, we compare three of the investigated algorithms: Dolev's approximate Byzantine agreement, the approximate BVC, and the Brooks–Iyengar algorithm.

4.4.1 Approximate Byzantine Agreement vs. Approximate BVC

We first compare the approximate Byzantine agreement with the approximate BVC. Given a set of vector inputs, we run the approximate Byzantine agreement algorithm for each individual dimension.

Let $V = \{\mathbf{v}_i = \{v_{i1}, v_{i2}, ..., v_{id}\} : 1 \le i \le N\}$ denote a set of d-dimensional measurements, where v_{ij} is the reading of dimension j, we assume $V = g \cup f$, where g is the set of $N - \tau$ valid measurements and f is the set of τ faulty measurements.

According to the validity condition of the approximate Byzantine agreement, the output of dimension l, denote by $\tilde{v}_{ABA}(l)$, satisfies $\min_g v^g(l) \le \tilde{v}_{ABA}(l) \le \max_g v^g(l)$, where $\min_g v^g(l)$ and $\max_g v^g(l)$ are the minimum valid measurement and maximum valid measurement of the lth element of the input vectors. Given the same set of inputs, the output of the approximate BVC, denote by $\tilde{v}_{BVC}$, is within the convex hull defined by the valid readings, i.e., $\tilde{v}_{BVC} \in \mathscr{H}(g)$. We can show that it is not hard to construct a set of inputs such that the output of the approximate Byzantine agreement falls out of convex hull $\mathscr{H}(g)$.

An 2-dimensional example is illustrated in Fig. 4.3, where $N = 6$, $\tau = 1$. The five black spots compose the input set of measurements. The spots corresponding to l_1, l_6 and h_1, h_6 are respectively chosen from $\min_g v^g(l)$ and $\max_g v^g(l)$, which define the precision of the approximate Byzantine agreement, i.e., the green and yellow square area. The red point is the faulty input of the two algorithms. Convex hull $\mathscr{H}(g)$ is the yellow area in Fig. 4.3. It is clear that $\tilde{v}_{ABA}$ might fall out of $\mathscr{H}(g)$. This shows that the approximate BVC is more precise than the approximate Byzantine algorithm mainly due to the convexity of the precision of approximate BVC.

Fig. 4.3 An example of extend approximate Byzantine agreement to process vector inputs ($N = 6$, $\tau = 1$)

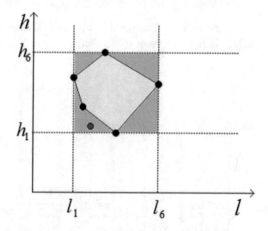

4.4.2 Approximate Byzantine Agreement vs. Brooks–Iyengar Algorithm

Approximate Byzantine agreement considers fusion of scalar readings while Brooks–Iyengar algorithm assumes measurements are intervals. To compare these two algorithms, we "convert" the interval measurements of Brooks–Iyengar algorithm into scalar values.

Consider a multiset of measurements $V = \{[l_1, h_1], [l_2, h_2], \cdots, [l_N, h_N]\}$, where $[l_i, h_i]$, $1 \leq i \leq N$ represents the reading received from PE_i. We calculate the midpoint of each measurement, denoted by $v_i = (l_i + h_i)/2$, for $1 \leq i \leq N$. And $v = [v_1, v_2, \ldots, v_N]$ is the scalarized counterpart of the V, and can be used as the input of approximate Byzantine agreement.

Proposition 4.1 *Given a set of N interval measurements $V = g \cup f$, where g is the set of $N - \tau$ valid measurements and f is the set of τ faulty measurements. If approximate Byzantine agreement takes the scalarized counterpart of V as input, there exist sets of interval measurements such that the output of approximate Byzantine agreement $\tilde{v}_{ABA} \notin [a^g_{N-2\tau}, b^g_{N-2\tau}]$, where $[a^g_{N-2\tau}, b^g_{N-2\tau}]$ is the region in WRD^g consisting of consecutive intervals with weights no smaller than $N - 2\tau$.*

Proof We assume that the midpoints of each measurement satisfy $v_1 \leq v_2 \leq \cdots \leq v_N$. In approximate Byzantine agreement, both the τ smallest and the τ largest readings are removed. Now we consider the $(\tau + 1)$th smallest value $v_{\tau+1}$.

If $v_{\tau+1}$ is a valid reading, assuming that $v_1, \ldots, v_\tau$ are faulty readings, then there are τ measurements with left endpoints smaller than $v_{\tau+1}$. $\tilde{v}_{ABA} < a^g_{N-2\tau}$ may occur when the number of interval measurements that contain $\tilde{v}_{ABA}$ is smaller than $N - 2\tau$.

By symmetry, we can prove that $\tilde{v}_{ABA}$ can be bigger than $b_{N-2\tau}$.

An example is illustrated in Fig. 4.4(1), where $a_{N-2\cdot1} = l_{2,i}$ and $\tilde{v}_{ABA} = (v_4 + v_2)/2$. If $l_{2,i} > (v_4 + v_2)/2$ then $\tilde{v}_{ABA} < a_2$. This happens mainly because that

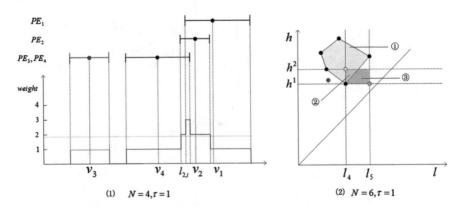

(1) $N = 4, \tau = 1$ (2) $N = 6, \tau = 1$

Fig. 4.4 Use scalars and vectors to represent intervals

we lose the bound information by only using midpoints instead of intervals. Also, the point estimation of Brooks–Iyengar algorithm is bounded by $[a_{N-2\tau}^g, b_{N-2\tau}^g]$, which is more close to true value.

4.4.3   Approximate BVC vs. Brooks–Iyengar Algorithm

Now we compare approximate BVC with Brooks–Iyengar algorithm. To apply BVC to a set of interval measurements V, we convert each interval $[l, h]$ to a 2-dimensional vector $[l\ h]^T$. In turn, given the output vector of approximate BVC, we can convert it back to an interval.

Proposition 4.2 *There exist sets of interval measurements such that the fusion output using approximate BVC is out of the accuracy bound of Brooks–Iyengar algorithm.*

Given a set of valid interval measurements $g = \{[l_1, h_1], \cdots, [l_{N-\tau}, h_{N-\tau}]\}$, assuming that $l_1 \leq l_2 \leq \cdots \leq l_{N-\tau}$, we then sort the right endpoints of each measurement in descending order $h^1 \leq h^2 \leq \cdots \leq h^{N-\tau}$. Note that h_i and h^i are not necessarily the same. The accuracy bound of Brooks–Iyengar algorithms lies in range $[l_\tau, h^{N-2\tau}]$. We can construct a set of interval measurements, the fusion output of which using is BVC is not within $[l_\tau, h^{N-2\tau}]$.

We give an example in Fig. 4.4(2), where the yellow point is the worst output vector $[a_{N-2\tau}\ b_{N-2\tau}]^T = [l_4\ h^2]^T$ of Brooks–Iyengar algorithm; the blue point is $[a_{N-\tau}\ b_{N-\tau}]^T = [l_5\ h^1]^T$; area ① and area ② are the convex hull of approximate BVC; area ② and area ③ are all the real output vectors (converted from interval that must contains $[a_{N-\tau}^g, b_{N-\tau}^g]$) of Brooks–Iyengar algorithm. The red point is the faulty input vector. Obviously, approximate BVC has could be worse than Brooks–Iyengar algorithm. This is because the fact that the left endpoints are always smaller than the right endpoints is never used in approximate BVC.

The comparison results show that the bound of Brooks–Iyengar algorithm is narrower than the bound of approximate Byzantine agreement and the bound of approximate BVC in terms of accuracy when they take interval measurements as input. If point values (approximate Byzantine agreement) or vectors (approximate BVC) are used, the useful information in the order relation of two endpoints of intervals no longer exist.

4.5   Precision Bound of the Brooks–Iyengar Algorithm

In this section, we analyze the precision bound of the Brooks–Iyengar algorithm. Symbols and terms used in the proof are listed below. Some are already defined in Table 4.1.

- g: A set of $N - \tau$ valid measurements.
- f: A set of τ faulty measurements.
- G: The set of all possible valid measurements, so $g \in G$.
- F: The set of all possible faulty measurements, so $f \in F$.
- $V = g \cup f$: A complete set of N measurements.
- WRD*: The weighted region diagram built using $*$, where $*$ is a set of measurements.
- Interval $I_w = [p_w, q_w]$: A continuous range in WRD overlapped by w measurements.
- Weight: The number of overlapped measurements of an interval.
- Region $[a_w^g, b_w^g]$: A continuous range in WRD consisting of consecutive intervals with weights no smaller than w.

4.5.1   Analysis and Proof of Precision Bound

Let ε_{BY} denote the precision of Brooks–Iyengar algorithm for $\forall g \in G$, which is the largest possible disagreement about the fusion output between two PEs:

$$\varepsilon_{BY} = \max_{i,j} |\tilde{v}_i - \tilde{v}_j| \qquad (4.18)$$

Since faulty sensors can send different measurements to different PEs and each PE performs data fusion independently, the calculation of ε_{BY} can be divided to maximizing $\tilde{v}_i$ and minimizing $\tilde{v}_j$ separately:

$$\varepsilon_{BY} = \max_i \tilde{v}_i - \min_j \tilde{v}_j, \ \forall g \in G \qquad (4.19)$$

This also allows us to calculate the bound of ε_{BY} by just considering one end of valid measurements (right end for maximization and left end for minimization). The result of the other end can be obtained by symmetry.

Theorem 4.5 *Consider a network of N sensors (PEs), at most $\tau = \lfloor N/3 \rfloor$ out of which are faulty ones. The precision bound of the Brooks–Iyengar algorithm is*

$$\frac{1}{1 + \alpha}(b_{N-2\tau}^g - a_{N-2\tau}^g) \qquad (4.20)$$

where $\alpha = \dfrac{N - \tau}{(2N - \tau)\tau}$.

Proof We start with maximizing $\tilde{v}_{BY}$

At any PE, for a faulty input to be able to affect the fusion output $\tilde{v}$, it must intersect with at least $N - 2\tau$ valid readings. According to Marzullo's algorithm, all valid readings must contain the correct value $\hat{v}$. Let WRD^g denote the weighted region diagram built from g, we have $[a_{w+1}^g, b_{w+1}^g] \subseteq [a_w^g, b_w^g]$. Therefore, faulty inputs must intersect with $[a_{N-2\tau}^g, b_{N-2\tau}^g]$ to be included in the calculation of $\tilde{v}$.

For $\forall f \in F$, let WRD^V be the weighted region diagram built using $V = g \cup f$. In this proof, intervals in WRD^V with weight no smaller than $N - \tau$ are called *effective intervals*. It is obvious that all effective intervals are within region $[a_{N-2\tau}^g, b_{N-2\tau}^g]$.

We rewrite (4.11) in a more general form

$$\tilde{v} = \frac{\sum_w w \cdot M_w}{\sum w}, \quad N - \tau \le w \le N \tag{4.21}$$

where M_w is the midpoint of an effective interval of weight w. Based on (4.21), the following are true and can be easily proved

1.

$$\min_w M_w \le \tilde{v} \le \max_w M_w \tag{4.22}$$

2. If $M_{w'} > \dfrac{\sum_w w \cdot M_w}{\sum w}$, then

$$\frac{\sum_w w \cdot M_w + w' \cdot M_{w'}}{\sum w + w'} > \frac{\sum_w w \cdot M_w}{\sum w} \tag{4.23}$$

3. If $w_1 > w_2$ and $M_{w_1} = M_{w_2}$, then

$$\frac{\sum_w w \cdot M_w + w_1 \cdot M_{w_1}}{\sum w + w_1} > \frac{\sum_w w \cdot M_w + w_2 \cdot M_{w_2}}{\sum w + w_2} \tag{4.24}$$

Since $\max_w M_w \le b_{N-2\tau}^g$, according to (4.22), we have

$$\tilde{v} < b_{N-2\tau}^g \tag{4.25}$$

So maximizing $\tilde{v}$ can be rephrased as making $\tilde{v}$ as close to $b_{N-2\tau}^g$ as possible. According to (4.22), (4.23) and (4.24), this means to have as many high-weight effective intervals as possible, with midpoint "M_w" as close to $b_{N-2\tau}^g$ as possible.

We first maximize the number of effective intervals. There is no doubt that all faulty readings should have their right ends no smaller than $b_{N-2\tau}^g$ for maximizing $\tilde{v}$. To maximize the number of intervals within $[a_{N-2\tau}^g, b_{N-2\tau}^g]$, the right ends of valid readings $(b_{N-\tau}^g, b_{N-\tau-1}^g, \cdots, b_{N-2\tau+1}^g)$ and the left ends of τ faulty readings are staggered. Because the number of intervals depends on the number of different endpoints within the region. And $N - \tau - 1$ different endpoints give $N - \tau$ intervals.

This strategy of staggering the endpoints could maximize $\sum_w w$ is not hard to prove by mathematical induction.

To obtain I_N, the highest-weight effective interval, all faulty measurements must intersect with $[a^g_{N-\tau}, b^g_{N-\tau}]$. This produces two distinct effective intervals of each $w \in [N-1, N-\tau]$, which is the maximum number of distinct intervals of the same weight as long as the right ends of faulty readings are no smaller than $b^g_{N-2\tau}$.

So far, we have maximized the number of effective intervals, as well as their weights. The last step is to maximize M_w for each effective interval. Let $\epsilon > 0$ be an infinitely small quantity, we define that

$$b^g_{N-2\tau} - b^g_{N-2\tau+1} = b^g_{N-2\tau+1} - b^g_{N-2\tau+2} = \cdots = b^g_{N-\tau-1} - b^g_{N-\tau} = \epsilon \qquad (4.26)$$

Let $l_1 < l_2 < \cdots < l_\tau$ be the left ends of the faulty readings, we also define that

$$l_1 - l_2 = l_2 - l_3 = \cdots = l_\tau - b^g_N = \epsilon \qquad (4.27)$$

From (4.26) and (4.27), we obtain

$$b^g_{N-2\tau} - l_1 = 2\tau \cdot \epsilon \qquad (4.28)$$

The multiple of an infinitely small positive is still an infinitely small positive. This guarantees any effective interval within $[l_1, b^g_{N-2\tau}]$ has a midpoint infinitely close to $b^g_{N-2\tau}$. It is possible to have $M_w = b^g_{N-2\tau}$, but this will reduce the number of effective intervals. Given that M_w is already infinitely close to $b^g_{N-2\tau}$, the number of effective intervals has much more influence over $\tilde{v}$ than M_w. In other words, increasing M_w from infinitely close to $b^g_{N-2\tau}$ to $b^g_{N-2\tau}$ at the cost of decreased number of intervals will actually reduce $\tilde{v}$. So to maximize $\tilde{v}$, ϵ should be as close to 0 as possible but never equal to 0.

A simple example of maximizing $\tilde{v}$ is illustrated in Fig. 4.5, where $N = 7, \tau = 2$. As shown in the figure, five effective intervals (two of $w = 5$, two of $w = 6$ and one of $w = 1$) are created with three valid readings (v_1, v_2, v_3) and two faulty readings (v_6, v_7). The width of each effective interval is ϵ, except for $I_5 = [a^g_5, b^g_3 - 2\epsilon]$.

From Fig. 4.5, we can see that the two effective intervals of the same weight $w \in [N-1, N-\tau]$ are symmetrical about the midpoint of I^V_N. This observation allows us to write $\tilde{v}_{\max}$ as

$$\frac{2(N-1)M_N + \cdots + 2(N-\tau+1)M_N + N \cdot M_N + (N-\tau)(M^l_{N-\tau} + M^r_{N-\tau})}{N + 2(N-1) + \cdots + 2(N-\tau)} \qquad (4.29)$$

where $M_N = b^g_{N-2\tau} - \frac{2\tau+1}{2}\epsilon$; $M^l_{N-\tau} = \frac{a^g_{N-\tau} + b^g_{N-2\tau} - 2\tau\epsilon}{2}$ is the midpoint of the left effective interval $I^l_{N-\tau} = [a^g_{N-\tau}, b^g_{N-\tau} - \tau\epsilon]$; and $M^r_{N-\tau} = b^g_{N-2\tau} - \frac{\epsilon}{2}$ is the midpoint of the right effective interval $I^r_{N-\tau} = [b^g_{N-2\tau} - \epsilon, b^g_{N-2\tau}]$. Simplifying (4.29) yields

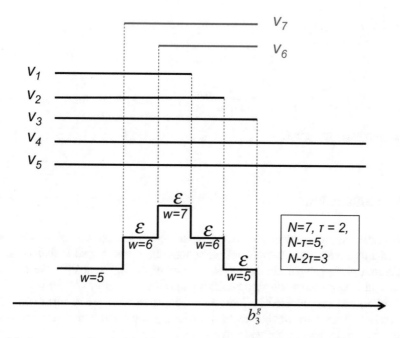

Fig. 4.5 An example of maximizing $\tilde{v}$

$$\tilde{v}_{max} = \frac{M_N(2N\tau - \tau^2 - N + \tau) + (N - \tau)(M_{N-\tau}^l + M_{N-\tau}^r)}{2N + 2N\tau - \tau - \tau^2} \quad (4.30)$$

By symmetry, we can write

$$\tilde{v}_{min} = \frac{\overline{M}_N(2N\tau - \tau^2 - N + \tau) + (N - \tau)(\overline{M}_{N-\tau}^l + \overline{M}_{N-\tau}^r)}{2N + 2N\tau - \tau - \tau^2} \quad (4.31)$$

where $\overline{M}_N = a_{N-2\tau}^g + \frac{2\tau+1}{2}\epsilon$, $\overline{M}_{N-\tau}^l = a_{N-2\tau}^g + \frac{\epsilon}{2}$ and $\overline{M}_{N-\tau}^r = \frac{a_{N-2\tau}^g + b_{N-\tau}^g + 2\tau\epsilon}{2}$.
Therefore, the precision bound of Brooks–Iyengar algorithm is given by

$$\lim_{\epsilon \to 0}(\tilde{v}_{max} - \tilde{v}_{min})$$

$$= \frac{(b_{N-2\tau}^g - a_{N-2\tau}^g)(2n\tau - \tau^2 - N + \tau) + (N - \tau)(b_{N-2\tau}^g - a_{N-2\tau}^g)}{N + 2N\tau - \tau^2 - \tau}$$

$$= \frac{2N\tau - \tau^2}{N + 2N\tau - \tau^2 - \tau}(b_{N-2\tau}^g - a_{N-2\tau}^g)$$

$$= \frac{1}{1 + \dfrac{N - \tau}{\tau(2N - \tau)}}(b_{N-2\tau}^g - a_{N-2\tau}^g) \quad (4.32)$$

If we let $\alpha = \frac{N-\tau}{\tau(2N-\tau)}$, (4.32) can be rewritten as

$$\lim_{\epsilon \to 0} (\tilde{v}_{max} - \tilde{v}_{min})$$

$$= \frac{1}{1+\alpha}(b^g_{N-2\tau} - a^g_{N-2\tau}) \tag{4.33}$$

This completes the proof.

4.6   Conclusion

This paper surveys a set of distributed, agreement-based sensor fusion algorithms and mainly investigates their precision bounds. Precision bound is defined as the maximum disagreement among fusion outputs of the different PEs. We focus on point and interval-based distributed fusion algorithms. Features of the investigated algorithms are summarized in Table 4.3. Compared with the summary in [52], we include more performance metrics, including precision bound, the maximum number of faulty PEs each algorithm can tolerate,[4] etc.

Table 4.3 Comparison of agreement-based distributed sensing algorithms

Algorithm	Approximate Byzantine agreement	FCA	Approximate BVC	Marzullo fusion algorithm	Brooks– Iyengar algorithm
Input	Scalar	Scalar	Vector	Interval	Interval
Faulty PEs tolerated	$< \frac{N}{3}$	$< \frac{N}{3}$	$\leq \frac{N-1}{d+2}$	$< \frac{N}{2}$	$< \frac{N}{2}$
Maximum faulty PEs	$< \frac{N}{3}$	$< \frac{2N}{3}$	$\leq \frac{N-1}{d+2}$	$< \frac{N}{2}$	$< \frac{2N}{3}$
Convergence rate	$1/(1 + \lfloor N -2\tau - 1 \rfloor)$	$\frac{2\tau}{N}$	$1-\gamma$	N/A	N/A
Accuracy bound	$\delta(g)$	$\frac{\kappa + \delta\tau}{N}$	Convex hull of non-faulty inputs	$[a^V_{N-\tau}, b^V_{N-\tau}]$	$[a^V_{N-\tau}, b^V_{N-\tau}]$
Precision bound	$\frac{\delta(g)}{2}$	$\frac{2\tau\delta}{N}$	$(1 - \gamma)(\Omega_l[t - 1] - \mu_l[t-1])$	$[a^g_{N-2\tau}, b^g_{N-2\tau}]$	$\frac{1}{1+\alpha}(b^g_{N-2\tau} - a^g_{N-2\tau})$ where $\alpha = \frac{N-\tau}{\tau(2N-\tau)}$

---

[4]The maximum faulty PEs of Brooks–Iyengar between [52] and this paper is different, since the definition is more rigid in this paper. In [52], it is whether the output interval contains the true value while it is whether the output is bounded by non-faulty inputs in this paper.

As shown in Table 4.3,

- *Fault tolerance*: FCA, Marzullo's algorithm, and Brooks–Iyengar algorithm may tolerate more faulty PEs than other algorithms.
- *Accuracy bound*: Approximate Byzantine agreement algorithm restricts output within the bound given by valid inputs. FCA cannot guarantee the output be better than the input in the worst-case. According to our definition of accuracy bound for interval-based fusion algorithms, Marzullo's sensor fusion method and Brooks–Iyengar algorithm provide the same accuracy level.
- *Precision bound*: All algorithms can iterate to improve the precision except Marzullo's method. For interval-based sensor fusion, Brooks–Iyengar algorithm provides better precision bound than the Marzullo's method.
- *Output format*: Only Brooks–Iyengar algorithm outputs both a point estimate and an interval estimate in one round. The Brooks–Iyengar algorithm considers different numbers of PEs and provides a balance between point estimation and fault tolerance, which can be extended to solve problems in other areas [52] (e.g., floating-point computations, software reliability).

Agreement-based distributed fusion plays an important role in distributed computing, sensor networks, and other distributed applications. Our work helps users select the proper fusion algorithm based on their features.

Much work remains to be done. The results in this paper provide essentially basic information on the ability of systems to tolerate errors and malicious faults. Architectural issues remain as to how these concepts can be integrated into Internet-scale architectures and large complex engineering systems. For examples, modern automobiles and aircraft are both complex networks of embedded systems. Fault detection and correction are essential. Much work remains to be done in combining these concepts with Internet systems and complex engineering systems with real-time feedback. There is a pressing need to create larger scale systems capable of cleansing themselves from malicious inputs.

Appendix: Proof of Theorem 4.3—Precision Bound of BVC

The following proof is based on the proof sketch in [27]. The proof of precision bound is based on that Z_i and Z_j both contain one identical point. Suppose that $m = N - \tau$ PEs are non-faulty and $\mathbf{v}_i[t]$, $\mathbf{v}_j[t]$ are estimate vectors of two non-faulty PEs at round t. In [27], Observations 1 and 3 in Part III of the proof of Theorem 4.5 imply that:

$$\mathbf{v}_i[t] = \sum_{k=1}^{m} \alpha_k \mathbf{v}_k[t-1] \qquad (4.34)$$

where $\sum_{k=1}^{m} \alpha_k = 1$, $\alpha_k \geq 0$, and

$$\mathbf{v}_j[t] = \sum_{k=1}^{m} \beta_k v_k[t-1] \tag{4.35}$$

where $\sum_{k=1}^{m} \beta_k = 1$, $\beta_k \geq 0$. Let g denote the index that satisfies $\alpha_g \geq \gamma$ and $\alpha_g \geq \gamma$. The existence proof of g is in [27], where $\gamma = 1/NC_{N-\tau}^{N}$.

$$v_{il}[t] = \sum_{k=1}^{m} \alpha_k v_{kl}[t-1]$$

$$\leq \alpha_g v_{gl}[t-1] + (1 - \alpha_g)\Omega_l[t-1]$$

$$\tag{4.36}$$

Since $v_{kl}[t-1] \leq \Omega_l[t-1]$, $\forall k$, (4.36) can be written as

$$v_{il}[t] \leq \gamma v_{gl}[t-1] + (\alpha_g - \gamma)v_{gl}[t-1] + (1-\alpha_g)\Omega_l[t-1]$$

$$\leq \gamma v_{gl}[t-1] + (\alpha_g - \gamma)\Omega_l[t-1] + (1-\alpha_g)\Omega_l[t-1]$$

$$\leq \gamma v_{gl}[t-1] + (1-\gamma)\Omega_l[t-1]$$

$$\tag{4.37}$$

Similarly, we obtain

$$v_{jl}[t] = \sum_{k=1}^{m} \beta_k v_{kl}[t-1]$$

$$\geq \beta_g v_{gl}[t-1] + (1 - \beta_g)\mu_l[t-1]$$

$$\geq \gamma v_{gl}[t-1] + (\beta_g - \gamma)v_{gl}[t-1] + (1-\beta_g)\mu_l[t-1]$$

$$\geq \gamma v_{gl}[t-1] + (\beta_g - \gamma)\mu_l[t-1] + (1-\beta_g)\mu_l[t-1]$$

$$\geq \gamma v_{gl}[t-1] + (1-\gamma)\mu_l[t-1]$$

$$\tag{4.38}$$

Subtracting (4.37) from (4.38) yields

$$\left| v_{il}[t] - v_{jl}[t] \right| \leq (1-\gamma)(\Omega_l[t-1] - \mu_l[t-1]) \tag{4.39}$$

References

1. S.R. Mahaney, F.B. Schneider, Inexact agreement: accuracy, precision, and graceful degradation, in *Proceedings of the Fourth Annual ACM Symposium on Principles of Distributed Computing (PODC)* (1985), pp. 237–249
2. D. Dolev, C. Dwork, L. Stockmeyer, On the minimal synchronism needed for distributed consensus. J. ACM **34**(1), 77–97 (1987)
3. M.H. Amini (ed.), in *Optimization, Learning, and Control for Interdependent Complex Networks*. Advances in Intelligent Systems and Computing, vol. 2 (Springer, Cham, 2020)
4. M.H. Amini, Distributed computational methods for control and optimization of power distribution networks, PhD Dissertation, Carnegie Mellon University, 2019
5. R.R. Brooks, S.S Iyengar, *Multi-sensor Fusion: Fundamentals and Applications with Software* (Prentice-Hall, Upper Saddle River, 1998)
6. L. Lamport, R. Shostak, M. Pease, The Byzantine generals problem. ACM Trans. Program. Lang. Syst. **4**(3), 382–401 (1982)
7. E.F. Nakamura, A.A.F. Loureiro, A.C. Frery, Information fusion for wireless sensor networks: methods, models, and classifications. ACM Comput. Surv. (CSUR) **39**(3), 9 (2007)
8. S.S. Blackman, T.J. Broida, Multiple sensor data association and fusion in aerospace applications. J. Robot. Syst. **7**(3), 445–485 (1990)
9. M. Barborak, A. Dahbura, M. Malek, The consensus problem in fault-tolerant computing. ACM Comput. Surv. (CSUR) **25**(2), 171–220 (1993)
10. R.C. Luo, M.G. Kay, Multisensor integration and fusion in intelligent systems. IEEE Trans. Syst. Man Cybern. **19**(5), 901–931 (1989)
11. F. Koushanfar, M. Potkonjak, A. Sangiovanni-Vincentell, Fault tolerance techniques for wireless ad hoc sensor networks, in *Sensors, 2002. Proceedings of IEEE*, vol. 2 (IEEE, Piscataway, 2002), pp. 1491–1496
12. T. Clouqueur, K.K. Saluja, P. Ramanathan, Fault tolerance in collaborative sensor networks for target detection. IEEE Trans. Comput. **53**(3), 320–333 (2004)
13. X. Luo, M. Dong, Y. Huang, On distributed fault-tolerant detection in wireless sensor networks. IEEE Trans. Comput. **55**(1), 58–70 (2006)
14. S.S. Iyengar, *Scalable Infrastructure for Distributed Sensor Networks* (Springer Science & Business Media, Berlin, 2006)
15. S.S. Iyengar, R.R. Brooks, *Distributed Sensor Networks: Sensor Networking and Applications* (CRC Press, Boca Raton, 2012)
16. M. Pease, R. Shostak, L. Lamport, Reaching agreement in the presence of faults. J. ACM **27**(2), 228–234 (1980)
17. M.J. Fischer, N.A. Lynch, M.S. Paterson, Impossibility of distributed consensus with one faulty process. J. ACM **32**(2), 374–382 (1985)
18. A.D. Fekete, Asymptotically optimal algorithms for approximate agreement, in *Proceedings of the Fifth Annual ACM Symposium on Principles of Distributed Computing (PODC)* (1986), pp. 73–87
19. I. Abraham, Y. Amit, D. Dolev, Optimal resilience asynchronous approximate agreement, in *8th International Conference on Principles of Distributed Systems, OPODIS 2004* (2004), pp. 229–239
20. D. Dolev, N.A. Lynch, S.S. Pinter, E.W. Stark, W.E. Weihl, Reaching approximate agreement in the presence of faults. J. ACM **33**(3), 499–516 (1986)
21. N.H. Vaidya, L. Tseng, G. Liang, Iterative approximate Byzantine consensus in arbitrary directed graphs, in *Proceedings of the 2012 ACM symposium on Principles of Distributed Computing* (ACM, New York, 2012), pp. 365–374
22. L. Tseng, N.H. Vaidya, Iterative approximate Byzantine consensus under a generalized fault model, in *Distributed Computing and Networking, 14th International Conference, ICDCN 2013. Proceedings* (2013), pp. 72–86

23. L. Su, N. Vaidya, Reaching approximate Byzantine consensus with multi-hop communication, in *Stabilization, Safety, and Security of Distributed Systems – 17th International Symposium, SSS 2015, Proceedings* (2015), pp. 21–35

24. B. Charron-Bost, M. Függer, T. Nowak, Approximate consensus in highly dynamic networks: the role of averaging algorithms, in *Automata, Languages, and Programming – 42nd International Colloquium, ICALP 2015, Proceedings, Part II* (2015), pp. 528–539

25. C. Li, M. Hurfin, Y. Wang, Brief announcement: reaching approximate Byzantine consensus in partially-connected mobile networks, in *Distributed Computing – 26th International Symposium, DISC 2012. Proceedings* (2012), pp. 405–406

26. C. Li, M. Hurfin, Y. Wang, Approximate Byzantine consensus in sparse, mobile ad-hoc networks. J. Parallel Distrib. Comput. **74**(9), 2860–2871 (2014)

27. N.H. Vaidya, V.K. Garg, Byzantine vector consensus in complete graphs, in *PODC* (2013), pp. 65–73

28. N.H. Vaidya, Iterative Byzantine vector consensus in incomplete graphs, in *Distributed Computing and Networking – 15th International Conference, ICDCN 2014. Proceedings* (2014), pp. 14–28

29. H. Mendes, M. Herlihy, Multidimensional approximate agreement in Byzantine asynchronous systems, in *Proceedings of the 45th Annual ACM Symposium on Theory of Computing (STOC)* (ACM, New York, 2013), pp. 391–400

30. A. Doudou, A. Schiper, Muteness detectors for consensus with Byzantine processes, in *Proceedings of the Seventeenth Annual ACM Symposium on Principles of Distributed Computing, PODC '98* (ACM, New York, 1998)

31. N.F. Neves, M. Correia, P. Verissimo, Solving vector consensus with a wormhole. IEEE Trans. Parallel Distrib. Syst. **16**(12), 1120–1131 (2005)

32. K. Marzullo, Tolerating failures of continuous-valued sensors. ACM Trans. Comput. Syst. **8**(4), 284–304 (1990)

33. K. Marzullo, S. Owicki, Maintaining the time in a distributed system, in *Proceedings of the Second Annual ACM Symposium on Principles of Distributed Computing*, pp. 295–305 (ACM, New York, 1983)

34. K.A. Marzullo, Maintaining the Time in a Distributed System: An Example of a Loosely-coupled Distributed Service (Synchronization, Fault-tolerance, Debugging). PhD thesis, Stanford, CA, USA, 1984. AAI8506272

35. S.S. Iyengar, D.N. Jayasimha, D. Nadig, A versatile architecture for the distributed sensor integration problem. IEEE Trans. Comput. **43**(2), 175–185 (1994)

36. D.N. Jayasimha, S. Sitharama Iyengar, R. Kashyap, Information integration and synchronization in distributed sensor networks. IEEE Trans. Syst. Man Cybern. **21**(5), 1032–1043 (1991)

37. D. Nadig, S.S. Iyengar, D.N. Jayasimha, A new architecture for distributed sensor integration, in *Southeastcon'93, Proceedings. IEEE* (IEEE, Piscataway, 1993), pp. 8–p

38. P. Blum, L. Meier, L. Thiele, Improved interval-based clock synchronization in sensor networks, in *Third International Symposium on Information Processing in Sensor Networks, 2004. IPSN 2004* (IEEE, Piscataway, 2004), pp. 349–358

39. L. Prasad, S. Sitharama Iyengar, R. Kashyap, R.N. Madan, Functional characterization of fault tolerant integration in distributed sensor networks. IEEE Trans. Syst. Man Cybern. **21**(5), 1082–1087 (1991)

40. S.S. Iyengar, L. Prasad, A general computational framework for distributed sensing and fault-tolerant sensor integration. IEEE Trans. Syst. Man Cybern. **25**(4), 643–650 (1995)

41. L. Lamport, Synchronizing time servers (1987). https://www.microsoft.com/en-us/research/uploads/prod/2016/12/Synchronizing-Time-Servers.pdf

42. U. Schmid, K. Schossmaier, How to reconcile fault-tolerant interval intersection with the Lipschitz condition. Distrib. Comput. **14**(2), 101–111 (2001)

43. D.N. Jayasimha, Fault tolerance in a multisensor environment, in *13th Symposium on Reliable Distributed Systems, 1994. Proceedings* (IEEE, Piscataway, 1994), pp. 2–11

44. P. Chew, K. Marzullo, Masking failures of multidimensional sensors, in in *Tenth Symposium on Reliable Distributed Systems, 1991. Proceedings* (IEEE, Piscataway, 1991), pp. 32–41

45. D. Desovski, Y. Liu, B. Cukic, Linear randomized voting algorithm for fault tolerant sensor fusion and the corresponding reliability model, in *Ninth IEEE International Symposium on High-Assurance Systems Engineering, 2005. HASE 2005* (IEEE, Piscataway, 2005), pp. 153–162
46. B. Parhami, Distributed interval voting with node failures of various types, in *IEEE International Parallel and Distributed Processing Symposium, 2007. IPDPS 2007* (IEEE, Piscataway, 2007), pp. 1–7
47. L. Prasad, S.S. Iyengar, R. Rao, R.L. Kashyap, Fault-tolerant integration of abstract sensor estimates using multiresolution decomposition, in *Proceedings of IEEE International Conference on Systems Man and Cybernetics*, vol. 5 (IEEE, Piscataway, 1993), pp. 171–176
48. L. Prasad, S.S. Iyengar, R.L. Rao, R.L. Kashyap, Fault-tolerant sensor integration using multiresolution decomposition. Phys. Rev. E **49**(4), 3452 (1994)
49. H. Qi, S. Sitharama Iyengar, K. Chakrabarty, Distributed multi-resolution data integration using mobile agents, in *Aerospace Conference, 2001, IEEE Proceedings*, vol. 3 (IEEE, Piscataway, 2001), pp. 3–1133
50. R. Brooks, S. Sitharama Iyengar, Optimal matching algorithm for multi-dimensional sensor readings, in *SPIE Proceedings of Sensor Fusion and Advanced Robotics, SPIE, Bellingham, Washington* (1995), pp. 91–99
51. R.R. Brooks, S. Sitharama Iyengar, Methods of approximate agreement for multisensor fusion, in *SPIE's 1995 Symposium on OE/Aerospace Sensing and Dual Use Photonics* (International Society for Optics and Photonics, Bellingham, 1995), pp. 37–44
52. R.R. Brooks, S. Sitharama Iyengar, Robust distributed computing and sensing algorithm. Computer **29**(6), 53–60 (1996)
53. V. Kumar, Impact of Brooks–Iyengar distributed sensing algorithm on real time systems. IEEE Trans. Parallel Distrib. Syst. **25**, 1370 (2013)
54. L. Xiao, S. Boyd, S. Lall, A scheme for robust distributed sensor fusion based on average consensus, in *Fourth International Symposium on Information Processing in Sensor Networks, 2005. IPSN 2005* (IEEE, Piscataway, 2005), pp. 63–70
55. R. Olfati-Saber, J.A. Fax, R.M. Murray, Consensus and cooperation in networked multi-agent systems. Proc. IEEE **95**(1), 215–233 (2007)
56. R. Olfati-Saber, R.M. Murray, Consensus problems in networks of agents with switching topology and time-delays. IEEE Trans. Autom. Control **49**(9), 1520–1533 (2004)
57. R.R. Murphy, Dempster-Shafer theory for sensor fusion in autonomous mobile robots. IEEE Trans. Robot. Autom. **14**(2), 197–206 (1998)
58. H. Wu, M. Siegel, R. Stiefelhagen, J. Yang, Sensor fusion using Dempster-Shafer theory [for context-aware HCI], in *Instrumentation and Measurement Technology Conference, 2002. IMTC/2002. Proceedings of the 19th IEEE*, vol. 1 (IEEE, Piscataway, 2002), pp. 7–12
59. V.V.S. Sarma, S. Raju, Multisensor data fusion and decision support for airborne target identification. IEEE Trans. Syst. Man Cybern. **21**(5), 1224–1230 (1991)

Chapter 5
The Profound Impact of the Brooks–Iyengar Algorithm

The Brooks–Iyengar algorithm has been inspiring to several researchers. One of the key researches that built upon this algorithm was developed by Dr. Gabriel Wainer, Associate Professor, Department of Systems and Computer Engineering. He wrote the following letter to Dr. Iyengar in 2012:

> During that period of research I was able to define new RT scheduling algorithms, and I included these (and other RT techniques) in the first existing RT version of an open-source OS (RT-Minix). The MINIX operating system was extended with real-time services, ranging from A/D drivers to new scheduling algorithms and statistics collection. (Gabriel Wainer, 1996, Carleton University)

5.1 Business, Media, and Academic References

In 1996, Iyengar's group, in collaboration with Brooks and with funding from Oak Ridge National Laboratory and the Office of Naval Research, invented a method of fault tolerance modeling that offers a computationally inspired real-time management solution. This work has emerged in new versions of real-time extensions to Linux operating systems. Many of these algorithms were used and installed in the RT-Linux operating system. They are now working on formal model verification by incorporating the algorithms into a new embedded kernel for robotic applications.

The profound contribution of the Brooks–Iyengar Distributed Computational Sensing work has enhanced new real-time features by adding fault tolerant capabilities.

In a paper entitled "Impact of Brooks–Iyengar Distributed Sensing Algorithm on Real Time Systems" published by IEEE Transactions on Parallel and Distributed Systems in 2014, V. Kumar evaluated the merit of Brooks–Iyengar distributed

© Springer Nature Switzerland AG 2020
P. Sniatala et al., *Fundamentals of Brooks–Iyengar Distributed Sensing Algorithm*,
https://doi.org/10.1007/978-3-030-33132-0_5

sensing algorithm and highlighted its impact on cost-effective processing of real-time sensor data stream. He discovered that it has a long-lasting impact, not only on sensor networks, but also on computer operating systems. He believed that as we learn more about the usefulness of this algorithm, it will become an essential component of systems that use sensors in some form.

Time to time, some algorithms appear and significantly affect technology. One such algorithm is Brooks–Iyengar Distributed Sensing Algorithm [1–5] that has had a profound impact on sensor technology similar to the effect the TCP/IP suite of protocols has had on data communication, Dijkstra's algorithm has had on process synchronization, and two-phase locking protocols has had on transaction serialization. It solved a number of complex issues in the deployment of large-scale sensor networks and continues to do so as the technology moves forward.

Dear Prof. Iyengar,

I am deeply honoured to have had the chance to discuss my past research on the Brooks–Iyengar algorithm on Robust Distributed Sensing and its application in Linux Operating Systems, as discussed on the phone last week. As I mentioned, in 1993 I was a researcher in the field of Real-Time Operating Systems (with a focus on Real-Time scheduling). During that period of research I was able to define new RT scheduling algorithms, and I included these (and other RT techniques) in the first existing RT version of an open-source OS (RT-Minix). The MINIX operating system was extended with real-time services, ranging from A/D drivers to new scheduling algorithms and statistics collection. A testbed was constructed to tests several sensor replication techniques in order to implement and verify several robust sensing algorithms. As a result, new services enhancing fault tolerance for replicated sensors were also provided within the kernel. The resulting OS offers new features such as real-time task management and clock resolution handling, and sensor replication manipulate Using this workbench, we implemented different versions of the Brooks–Iyengar algorithm for robust sensing, using inexact agreement and optimal region. The introduction of this new mechanism provided more accuracy and precision.

These results were published in various papers and a book. My ideas were used shortly after by other researchers in the field, leading to the development of the first versions of RT-Linux. Fifteen years after, this approach continues to be used and cited, and new Real-Time projects based on the concepts I defined have been started in the last 5 years. These results were published in various papers and a book. My ideas were used shortly after by other researchers in the field, leading to the development of the first versions of RT-Linux. Fifteen years after, this approach continues to be used and cited, and new Real-Time projects based on the concepts I defined have been started in the last 5 years.

Sincerely,

Gabriel Wainer.

5.2   Industrial Projects Incorporating the Algorithm

Carleton University is developing a formal model verification by incorporating the algorithm into a new embedded kernel for robotic applications. Dr. Gabriel Wainer, Associate Professor at Carleton University in 2012 wrote the following letter to Dr. Iyengar:

   Also, in April 2013, Raytheon Company wrote the following letter in regards to the profound contribution of the algorithm in industry:

Dear Prof. Iyengar,

I am deeply honoured to have had the chance to discuss my past research on the Brooks–Iyengar algorithm on Robust Distributed Sensing and its application in Linux Operating Systems, as discussed on the phone last week. As I mentioned, in 1993 I was a researcher in the field of Real-Time Operating Systems (with a focus on Real-Time scheduling). During that period of research I was able to define new RT scheduling algorithms, and I included these (and other RT techniques) in the first existing RT version of an open-source OS (RT-Minix). The MINIX operating system was extended with real-time services, ranging from A/D drivers to new scheduling algorithms and statistics collection. A testbed was constructed to tests several sensor replication techniques in order to implement and verify several robust sensing algorithms. As a result, new services enhancing fault tolerance for replicated sensors were also provided within the kernel. The resulting OS offers new features such as real-time task management (for both periodic or a periodic tasks), clock resolution handling, and sensor replication manipulate Using this workbench, we implemented different versions of the Brooks–Iyengar algorithm for robust sensing, using inexact agreement and optimal region. The introduction of this new mechanism provided more accuracy and precision.

   These results were published in various papers and a book. My ideas were used shortly after by other researchers in the field, leading to the development of the first versions of RT-Linux. Fifteen years after, this approach continues to be used and cited, and new Real-Time projects based on the concepts I defined have been started in the last 5 years. These results were published in various papers and a book. My ideas were used shortly after by other researchers in the field, leading to the development of the first versions of RT-Linux. Fifteen years after, this approach continues to be used and cited, and new Real-Time projects based on the concepts I defined have been started in the last 5 years.

Sincerely,

Gabriel Wainer.

The Raytheon BBN Web Site states "Each Raytheon BBN Technologies breakthrough can have a massive impact on the way we live—a powerful motivator for both customers and employees." This is true of the Brooks–Iyengar Algorithm':

"The algorithm referred to as the Brooks–Iyengar algorithm or the Brooks–Iyengar hybrid algorithm is a distributed algorithm that improves both the precision and accuracy of the measurements taken by a distributed sensor network, even in the presence of faulty sensors," In 2002, Raytheon BBN was the Integration and Test Team lead for DARPA's Sensor Information Technology (SensIT) program under DARPA PM Dr. Sri Kumar.

To meet DARPA's goal of developing technology for distributed sensor processing, Raytheon BBN designed a sensor network architecture for which other SensIT participants developed components. A Penn State Applied Research Laboratory (ARL) team led by Dr. Richard Brooks developed and demonstrated the Brooks–Iyengar algorithm as a "collaborative application" in the reactive sensor nets (RSN) instance on the SensIT software architecture that integrated the Brooks–Iyengar algorithm with other work from University of Southern California Information Sciences Institute (USC/181), BAE Systems, Fantastic data, the University of Maryland, and others. This work was demonstrated at the Marine Corps Air Ground Combat Center, Twentynine Palms, California, and in a test network set up at BBN's (formerly GTE) Waltham, Massachusetts laboratory. This was Raytheon's initial experience with the Brooks–Iyengar algorithm in a realistic setting.

Since then, the impact of this work has been significant. The Brooks–Iyengar Algorithm makes it possible to extract reliable data from sensor networks when some sensors are unreliable, thus adding a fault tolerant aspect. Unreliable sensor data are typically the case in systems that Raytheon develops for its customers, so application of the algorithm made Raytheon and its customers more successful without increased investment in improved sensor reliability. This saved money, always important in today's increasingly budget conscious environment.

In summary, the Brooks–Iyengar algorithm continues to have significant impact where it is applied to Raytheon's programs.

"... The seminal work of professor Iyengar and his colleagues on sensor networks infrastructure constitutes a foundation for sensor placement and optimization of information obtained from sensor networks. The ideas and algorithms introduced by Iyengar et al. Played an important role in Telcordia patent 019576 entitled "Method for placement of sensors for Surveillance" and Motorola patent 7,688,793 entitled, "Wireless sensor node group affiliation method and apparatus," both of which build up upon his pioneering work. I believe that professor Iyengar work on sensor deployment, coverage, and surveillance has led to major advances in network technologies ...."

5.3 Impacts of Brooks–Iyengar Algorithms on Academic Dissertations

Introduction Iyengar's work has had a significant impact in industrial applications/research, and continuing scientific research. Many new ideas have been inspired by his seminal work, resulting in a variety of new innovations by industries and universities.

The intellectual advancement of his ideas has been cited in over 28 Ph.D. dissertations and other new thesis and dissertations resulting in scientific discoveries around the world. His work has impacted thesis in Germany, the USA, and China. His foundational work has also resulted in innumerable funding opportunities provided by the National Science Foundation to continue to advance research in this significant area.

More than 28 dissertations from highly ranked universities (e.g., MIT, Carnegie Melon University, RICE University, University of Southern California, and University of Pennsylvania) have extensively inspired by the Brooks–Iyengar Algorithm or cited it in their dissertation. Here are some examples:

Brooks Iyengar Algorithm An algorithm to exchange data with accuracy is the Brooks–Iyengar algorithm. The Brooks–Iyengar algorithm is a hybrid algorithm that combines data fusion with Byzantine agreement to filter out-of-range values and average the accepted values. The rough steps of the algorithm are as follows:

1. First the node shares the data to aggregate with its neighbors.
2. The second step is to filter out invalid (out of range) information, which results in the lower and upper bound and accuracy of the accepted values.
3. The final step is to calculate the weighted average of the accepted values, where the weights are the number of sensors whose readings intersect with each of the accepted values. An attacker can falsify the aggregated result, but it will not hurt the normal operation of the system.

5.3.1 The SDSN-Aggregation

Brooks and Iyengar give a high-level view of a multi-sensor tracking system, also shown in Fig. 5.1.

Each sensor node continuously monitors its environment and tries to detect events as they occur within the sensor's field of operation. Target information is then used by the node to create a detection event by matching it with the node's known target type database. Local collaboration is performed to create an accurate categorization of the target, and includes only nodes within a dynamically determined geographic neighborhood and time frame. Results of this local collaboration are then used to initiate and maintain tracking of the target. Tracking information

Fig. 5.1 High-level view of
the multi-sensor tracking
system

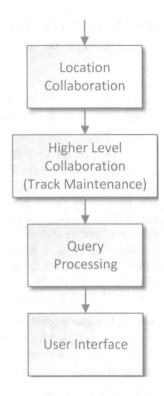

may also be stored in a distributed database which is possibly part of a complex
query processing system and is tied to a user interface.

As can be deduced, detection and in-node processing is the first step for target
tracking. Robustness and reliability are also important factors one must consider
in this process of sensor fusion. In this context, the number of sensor failures
the network can tolerate becomes crucial, as is the manner in which data from
fit sensors is separated from the unfit ones. Richard Brooks et al. give a solution
that satisfies the requirements of inexact agreement problem by merging the sensor
fusion algorithm with Mahaney and Schneider's fast convergence algorithm. The
solution assumes the self-organization of sensor nodes into clusters. The sensor
fusion algorithm runs on the cluster head, which collects the processed data from
the group members and inputs this into the algorithm.

A flowchart for the process of target tracking is shown in Fig. 5.2. The initialization phase consists of publish–subscribe calls. Brooks–Iyengar's methodology
proposes two waves of publish–subscribe calls propagated in four directions.

Once the necessary sensors are deployed on the ground, their data is transmitted
back to the base in order to provide decision makers with the necessary situational
awareness. In recent years researchers have given more attention to operational
issues of WSNs such as energy-efficient communication and data fusion [6].

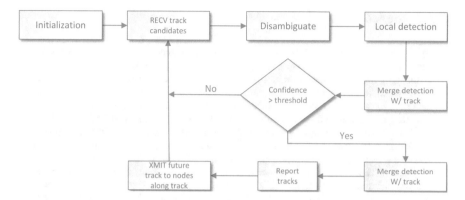

Fig. 5.2 High-level view of the multi-sensor target tracking process

Sensor redundancy has been explored in depth in the area of sensor fusion where sensors are generally considered to measure the same variable but through different means and with different accuracies. One of the first works in this field assumes that sensors provide one-dimensional intervals and shows worst-case results regarding the size of the fused interval based on the number of faulty sensors in the system. A variation of this work relaxes the worst-case guarantees in favor of obtaining more precise fused measurements through weighted majority voting [6].

In the medical area, more and more sophisticated sensors have been used in medical diagnosis. Sensors, such as nuclear magnetic resonance (NMR) devices, acoustic imaging devices, and fibreoptic probes, provide various ways to examine patients. If these devices and their data could be integrated, the accuracy and reliability of diagnosis would be greatly improved [6].

In recent years, data fusion has been increasingly applied to civilian applications including remote sensing, automated control of industrial manufacturing systems, medical diagnosis, and food quality and safety inspection [6].

Slow increase of the performance of single-processor caused by transmission speeds, limits to miniaturization and economic limitations 2 leads to designing microprocessors in the direction of parallelism. Parallel and distributed computing not only has been used to solve difficult problems in many areas of science and engineering such as communication, computing, and sensor network [6] but also attracts people's attention from industry for its ability to process large amount of data in sophisticated ways.

Fusion Fusion involves taking input from multiple ambiguous sources about the same input event (intended action by the user) and combining them to get one set of ambiguous possible actions. Although it is rare to have redundant sensors in the GUI world, it is quite common in a robust sensing environment. Thus, a complete solution to the problem of ambiguity in the context-aware computing world would benefit from structured support for the fusion problem, a problem that has been extensively addressed in other research areas such as robotics [6].

5.3.2   Sensor Fusion

Cyber-physical systems (CPS) are large, distributed embedded systems that integrate sensing, processing, networking, and actuation. Developing CPS applications is currently challenging due to the sheer complexity of the related functionality as well as the broad set of constraints and unknowns that must be tackled during operation. Building accurate data representations that model the behavior of the physical environment by establishing important data correlations and capturing physical laws of the monitored entities is critical for dependable decision-making under performance and resource constraints. The goal of this thesis is to produce reliable data models starting from raw sensor data under tight resource constraints of the execution platform while satisfying the timing constraints of the application. This objective was achieved through adaptation policy designs that optimally compute the utilization rates of the available network resources to satisfy the performance requirements of the application while tracking physical entities that can be quasistatic or dynamic in nature. The performance requirements are specified using a declarative, high-level specification notation that corresponds to timing, precision, and resource constraints of the application. Data model parameters are generated by solving differential equations using data sampled over time and modeling errors occur due to missed data correlations and distributed data lumping of the model parameters.

In this thesis, the Brooks–Iyengar algorithm is used two times: (1) as a platform to validate the proposed approach: An updated fragment structure which is compared with existing leader-based, cluster-based, and centralized techniques such as Brooks–Iyengar approach, (2) as a promising approach that defines a centralized threshold-OR fusion rule for combining sensor samples under normally distributed, independent additive noise conditions.

"Wireless Sensor Networks (WSNs) are rapidly emerging as a new class of distributed pervasive systems, with applications in a diverse range of domains such as traffic management, building environment management, target tracking, etc. Most, if not all, WSN application design is currently performed at the node-level, with developers manually customizing various protocols to realize their applications. This makes it difficult for the typical domain-expert application developer (e.g., a building system designer) to develop applications for them, and is a hindrance to their wide acceptance. To address this problem, the field of macroprogramming has emerged, which aims to provide high-level programming abstractions to assist in application development for WSNs. Although several macroprogramming approaches have been studied, the area of compilation of these macroprograms to node-level code is still largely unexplored. This thesis addresses the issues involved in the compilation of sensor network macroprograms. The emphasis is on data-driven macroprogramming, where the application is represented as a set of tasks running on the system's nodes—producing, processing, and acting on data items or streams to achieve the system's goals. In addition to a modular framework for the overall compilation process, formal models for the task-mapping problem which arises in this context are discussed. Results from optimal result-producing mixed-integer programming techniques and algorithmic heuristics for the above problem are presented. We also present the design and implementation of a graphical toolkit for sensor network macroprogramming." In this thesis, Brooks–Iyengar algorithm is introduced as an excellent approach that explores various aspects of networked sensing.

This PhD thesis is inspired by Brooks–Iyengar algorithm. According to this thesis, they characterize, analytically and quantitatively, the bit error rate (BER) process, the network lifetime and the topological properties of wireless sensor networks (WSN) that was firstly developed by Iyengar and Brooks.

This thesis is inspired by Brooks–Iyengar algorithm. According to this thesis, they developed a system for unmanned aerial vehicle (UAV) swarm which is a heterogeneous network of mobile vehicles whose sensors collect data. "A graph is a key construct for expressing relationships among objects, such as the radio connectivity between nodes contained in an unmanned vehicle swarm. The study of such networks may include ranking nodes based on importance, for example, by applying the PageRank algorithm used in some search engines to order their query responses. The PageRank values correspond to a unique eigenvector typically computed by applying the power method, an iterative technique based on matrix multiplication. The first new result described herein is a lower bound on the execution time of the PageRank algorithm that is derived by applying standard assumptions to the scaling value and numerical precision used to determine the PageRank vector. The lower bound on the PageRank algorithm's execution time also equals the time needed to compute the coarsest equitable partition, where that partition is the basis of all other results described herein. The second result establishes that nodes contained in the same block of a coarsest equitable partition must yield equal PageRank values. The third result is an algorithm that eliminates differences in the PageRank values of nodes contained in the same block if the PageRank values are computed using finite-precision arithmetic. The fourth result is an algorithm that reduces the time needed to find the PageRank vector by eliminating certain dot products when any block in the partition contains multiple vertices. The fifth result is an algorithm that further reduces the time required to obtain the PageRank vector of such graphs by applying the quotient matrix induced by the coarsest equitable partition. Each algorithm's complexity is derived with respect to the number of blocks contained in the coarsest equitable partition and compared to the PageRank algorithm's complexity."

5.4 Algorithm Potential for Future Market Growth

Iyengar's work has had a significant impact in industrial applications/research, and continuing scientific research. Many new ideas have been spawned from his seminal work, resulting in a variety of new innovations by industries such as Motorola, Telcordia, Boeing, and even new chip-design at universities. New applications from industrial research are ongoing, each resulting in millions of dollars of market growth.

The intellectual advancement of his ideas has been incorporated into over 28 Ph.D. thesis and other new thesis and dissertations resulting in scientific discoveries around the world. His work has impacted thesis in Germany, the USA, and China. His foundational work has also resulted in innumerable funding opportunities provided by the National Science Foundation to continue to advance research in this significant area.

5.5   Related Contribution to Sensor Networks by S. S. Iyengar

5.5.1   Optimal Sensor Placement Algorithms

S. S. Iyengar is the co-inventor of an optimal sensor placement technique relating to a limited number of sensors at selected locations in order to provide protection to all locations in real-time applications, which has been included in patients by Telcordia and Motorola, as well as other companies (IEEE Computers, 2002[1]). This groundbreaking work, which has been cited more than 1000 times, has laid an intellectual foundation for designing policies and techniques in many areas of real-time applications from surveillance systems to telecommunications systems.

5.5.1.1   First in the Literature

Prior to this work, the literature of distributed sensor networking has largely ignored the above sensor placement problem. Most prior work had concentrated exclusively on efficient sensor communication and sensor ion for a given sensor field architecture. However, as sensors are used in greater numbers for field operation, efficient deployment strategies become increasingly important. Related work on terrain model acquisition for motion planning has focused on the movement of a robot in an unexplored "sensor field." While knowledge of the terrain is vital for surveillance, it does not directly solve the sensor placement problem.

5.5.1.2   Summary of Contribution

We formulate the q-coverage deployment problem as an ILP with $O(tn^2)$ variables and $O(tn^2)$ equations, where n is the number of grid points. For a large n, we propose a divide-and conquer "near-optimal" algorithm in which the base case (a small number of points) is solved optimally by using the ILP formulation.

5.5.1.3   Impact on PhD Dissertations

The intellectual advancement of his ideas has been incorporated into over 44 Ph.D. theses and other new theses and dissertations resulting in scientific discoveries around the world. His work has impacted thesis in Germany, the USA, and China (see Appendix 2 in this chapter for the complete list of Ph.D. dissertations).

---

[1]Grid Coverage for Surveillance and Target Location in Distributed Sensor Networks, IEEE Trans. on Computers Vol 51 No. 12 Dec. 2002.

5.5.1.4 Industry Impacts

Iyengar's work has had a significant impact in industrial applications/research, and continuing scientific research. Many new ideas have been spawned from his seminal work, resulting in a variety of new innovations by many industries and even new chip-design at universities. New applications from industrial research are ongoing, each resulting in millions of dollars of market growth.

5.6 Conclusion and Outlook

Sensors have become highly pervasive. They have become the spinal cord of every system. The Brooks–Iyengar hybrid algorithm for distributed control in the presence of noisy data combines Byzantine agreement with sensor fusion. It bridges the gap between sensor fusion and Byzantine fault tolerance. This seminal algorithm unified these disparate fields for the first time. Essentially, it combines Dolev's algorithm for approximate agreement with Mahaney and Schneider's fast convergence algorithm (FCA). The algorithm assumes N processing elements (PEs), t of which are faulty and can behave maliciously. It takes as input either real values with inherent inaccuracy or noise (which can be unknown), or a real value with a priori defined uncertainty, or an interval. The output of the algorithm is a real value with an explicitly specified accuracy. The algorithm runs in $O(N \log N)$ and can be modified to correspond to Crusader's Convergence Algorithm (CCA); however, the bandwidth requirement will also increase. The algorithm has applications in different fields from distributed control, software reliability, and High-performance computing. The struggle for efficient inter-sensor and intra-sensor communications has found its lifeblood in this algorithm. Take an example of driverless cars and automatic control of traffic that presents us with a highly complex distributed sensor network. It requires accurate data sharing for accident-free traffic control, self-synchronization [7, 8] of moving vehicles at road intersections, and a number of other communication tasks. The Brooks–Iyengar algorithm will play a crucial role in the automation of moving vehicles in 2D and 3D space. This chapter discussed the impact of Brooks–Iyengar algorithm on DARPA's program, real-time UNIX systems, industrial companies, and academic theses/dissertations. It also addressed the algorithm potentials for future market growth.

Appendix 1

This appendix lists Ph.D. dissertations/theses that have incorporated Brooks–Iyengar Algorithm or cited it in their research monographs:

1. Heuven, D. W. Opportunistic sensing & aggregation network using smart-phones and sensor nodes. MS thesis. **University of Twente**, 2014.

2. Ivanov, Radoslav Svetlozarov. Context-Aware Sensor Fusion for Securing Cyber-Physical Systems. Diss. **University of Pennsylvania**, 2017.
3. Venugopal, Subalakshmi. Rendezvous Reservation Protocol for Energy Constrained Wireless Infrastructure Networks: Impact of Battery Management Mechanisms. Diss. **Washington State University**, 2003.
4. Naik, Udayan. Implementation of distributed composition service for self-organizing sensor networks. Diss. **Auburn University** 2005.
5. Jourdan, Damien. Wireless sensor network planning with application to UWB localization in GPS-denied environments. Diss. **Massachusetts Institute of Technology**, 2006.
6. Li, Changying. Sensor fusion models to integrate electronic nose and surface acoustic wave sensor for apple quality evaluation. Diss. **Pennsylvania State University** 2006.
7. Mankoff, Jennifer. An Architecture and Interaction Techniques for Handling Ambiguity in Recognition-Based Input. Diss. **Carnegie Melon University** 2001.
8. Coopmans, Calvin. Cyber-physical systems enabled by unmanned aerial system-based personal remote sensing: Data mission quality-centric design architectures. **Utah State University**, 2014.
9. Dua, Sumeet. Techniques to explore time-related correlation in large datasets. Diss. **Louisiana State University**, 2002.
10. Lee, Hojun. Ontology-based data fusion within a net-centric information exchange framework. Diss. **University of Arizona**, 2009.
11. Reichle, Roland. Information Exchange and Fusion in Dynamic and Heterogeneous Distributed Environments. Diss. **University of Kassel**, 2010.
12. Atrey, Pradeep Kumar. Information assimilation in Multimedia surveillance systems. Diss. **National University of Singapore**, 2007.
13. Polina, Phani Chakravarthy. SDSF: social-networking trust based distributed data storage and co-operative information fusion. Diss. **University of Louisville**, 2014.
14. Bakhtiar, Qutub Ali. Mitigating inconsistencies by coupling data cleaning, filtering, and contextual data validation in wireless sensor networks. Diss. **Florida International University**, 2009.
15. Vemula, Mahesh. Monte Carlo methods for signal processing in wireless sensor networks. **State University of New York at Stony Brook**, 2007.
16. Subramanian, Varun. Building Distributed Data Models in a Performance-Optimized, Goal-Oriented Optimization Framework for Cyber-Physical Systems. Diss. **State University of New York at Stony Brook**, 2012.
17. Almasri, Marwah M. Multi Sensor Fusion Based Framework For Efficient Mobile Robot Collision Avoidance and Path Following System. Diss. **University of Bridgeport**, 2016.
18. Maslov, Igor V. Improving the performance of Evolutionary algorithms in imaging optimization. **City University of New York**, 2008.
19. Peng, Zhimin. Parallel sparse optimization. Diss. **Rice University**, 2013.

20. Herath, Vijitha Rohana. Stability of Distributed Sensor Networks. Diss. **University of Miami**, 2002.
21. Mysorewala, Muhammad Faizan. Simultaneous robot localization and mapping of parameterized spatio-temporal fields using multi-scale adaptive sampling. Diss. **The University of Texas at Arlington**, 2008.
22. Bell, Iverson C. A System Concept Study and Experimental Evaluation of Miniaturized Electrodynamic Tethers to Enhance Picosatellite and Femtosatellite Capabilities. Diss. **University of Michigan** 2015.
23. Chakrabarty, Nirupam. Semiparametric Estimation of Target Location in Wireless Sensor Network. Diss. **University of Michigan**, 2014.
24. Pathak, Animesh. Compilation of data-driven macroprograms for a class of networked sensing applications. **University of Southern California**, 2008.
25. Augeri, Christopher J. On graph isomorphism and the PageRank algorithm. **Air Force Institute of Technology**, 2008.
26. Guo, Xin. Occupancy sensor networks for improved lighting system control. The **University of Nebraska-Lincoln**, 2007.
27. Ilyas, Muhammad Usman. Analytical and quantitative characterization of wireless sensor networks. **Michigan State University**, 2009.
28. Bruck, Hugh A. New Metrological Techniques for Mechanical Characterization at The Microscale and Nanoscale. Diss. **University of Maryland**, 2004.

Appendix 2

This appendix lists Ph.D. dissertations/theses that have incorporated Iyengar's sensor placement algorithm or cited it in their research monographs:

1. Ghafoori, Amir. Decision analytics for sonar placement to mitigate maritime security risk. Rutgers The State University of New Jersey-New Brunswick, 2013.
2. Cashbaugh, Jasmine. "Cluster Control of a Multi-Robot Tracking Network and Tracking Geometry Optimization." (2016).
3. Ramsden, Daryn. Optimization approaches to sensor placement problems. Diss. Rensselaer Polytechnic Institute, 2009.
4. Kavalapara, Rahul. "Energy-Efficient Fault Tolerant Coverage for Wireless Sensor Networks." (2010).
5. Jourdan, Damien. Wireless sensor network planning with application to UWB localization in GPS-denied environments. Diss. Massachusetts Institute of Technology, 2006.
6. Huang, Meng-chiang. "Collaborative Detection of Unauthorized Traversals in Mobile Sensor Networks." (2014).
7. Khalil, Issa. "Mitigation of Control and data traffic attacks in wireless ad-hoc and sensor networks." Purdue University(2007).

8. Dong, Shaoqiang. Node Placement, Routing and Localization Algorithms for Heterogeneous Wireless Sensor Networks. Diss. 2008.
9. Yonga, Franck Ulrich Yonga. Enabling Runtime Self-Coordination of Reconfigurable Embedded Smart Cameras in Distributed Networks. University of Arkansas, 2015.
10. Dewasurendra, Duminda A. Evidence filtering and its distributed implementation on grid sensor networks. University of Notre Dame, 2009.
11. Seetharaman, Sumathi. Self-organized scheduling of node activity in large-scale sensor networks. Diss. University of Cincinnati, 2004.
12. Bagheri, Majid. Efficient k-coverage algorithms for wireless sensor networks and their applications to early detection of forest fires. Diss. School of Computing Science-Simon Fraser University, 2007.
13. Nagilla, Praveen Kumar. Sensor coverage and actors relocation in wireless sensor and actor networks (WSAN):l optimization models and approximation algorithms. Diss. University of Missouri–Columbia, 2010.
14. Vlasenko, Iuliia. Deployment planning for location recognition in the Smart-Condo: Simulation, empirical studies and sensor placement optimization. Diss. University of Alberta (Canada), 2013.
15. Marshall, Michael Brian. "A swarm intelligence approach to distributed mobile surveillance." (2005).
16. Zheng Shujun, Indoor Positioning System Using Pyroelectric Infrared Sensors for Wireless Sensing Networks." China, (2013).
17. Yedavalli, Kiran Kumar. On location support and one-hop data collection in wireless sensor networks. University of Southern California, 2007.
18. Al-Omari, Safwan. Petra: Toward dependable and autonomic networked sensor systems. Diss. Wayne State University, 2009.
19. Deyab, Tamer Mohamed. Optimization of Sensors Deployment in a 3D Environment under the Coverage, Connectivity and Energy Consumption Constraints. Diss. King Fahd University of Petroleum and Minerals (Saudi Arabia), 2011.
20. Yildiz, Enes. Providing multi-perspective coverage in wireless multimedia sensor networks. Southern Illinois University at Carbondale, 2011.
21. Eli, Haluk. Terrain visibility and guarding problems. Diss. Bilkent University, 2017.
22. El-Rewini, Hesham, et al. "ON THE DEPLOYMENT OF MOBILE HETERO-GENEOUS SENSORS IN CRITICAL INFRASTRUCTURE."
23. Wu, Xiaoling. Coverage-driven Energy-efficient Deployment and Self-organization in Sensor Networks. Diss. Thesis for the Degree of Doctor of Philosophy, Dept. Comput. Eng., Graduate School, Kyung Hee Univ., Seoul, Korea, 2008.
24. Watfa, Mohamed Khalil. Coverage issues in wireless sensor networks. Diss. 2006.
25. Xiang, Yun. "Mobile Sensor Network Design and Optimization for Air Quality Monitoring." (2014).

26. Ergen, Sinem Coleri. Wireless sensor networks: Energy efficiency, delay guarantee and fault tolerance. Diss. University of California, Berkeley, 2005.
27. Rocha, Bruno Filipe Ferreira Graa. Structural health monitoring of aircraft structures: development of a phased array system. Diss. 2010.
28. Ababnah, Ahmad A. Sensor deployment in detection networks-a control theoretic approach. Diss. Kansas State University, 2010.
29. Erande, Kaustubh Rajan. Design of a user driven real time asset tracking system using RFID in a healthcare environment. Oklahoma State University, 2008.
30. Das, Nibedita. Coverage and connectivity problems for sensor networks. Arizona State University, 2010.
31. Erickson, Lawrence H. Visibility analysis of landmark-based navigation. University of Illinois at Urbana-Champaign, 2014.
32. Soriente, Claudio. Data security in unattended wireless sensor networks. University of California, Irvine, 2009.
33. Zhao, Jian. Camera planning and fusion in a heterogeneous camera network. University of Kentucky, 2012.
34. Fusco, Giordano. Coverage Optimization in Sensor and Cellular Networks. Diss. State University of New York at Stony Brook, 2013.
35. Ramadan, Rabie A. On the deployment of mobile heterogeneous sensors in critical infrastructure. Diss. Southern Methodist University, 2007.
36. Sanli, Hidayet Ozgur. Energy aware node scheduling and multiple target tracking with event miss-ratio assurances in wireless sensor networks. Arizona State University, 2007.
37. Pandey, Santosh. Secure localization and node placement strategies for wireless networks. Auburn University, 2007.
38. Hays, Jacob. Control of Self-Reconfigurable Robot teams for sensor placement. Rochester Institute of Technology, 2010.
39. Khazaeni, Yasaman. An event-driven approach to control and optimization of multi-agent systems. Diss. Boston University, 2016.
40. Alnawaiseh, Ala. A modeling framework for the design and evaluation of cooperative wireless sensor networks. Diss. Southern Methodist University, 2012.
41. Cho, Shung Han. Collaborative and heterogeneous signal processing methodology for mobile sensor based applications. State University of New York at Stony Brook, 2010.
42. Golen, Erik F. Intelligent deployment strategies for passive underwater sensor networks. Rochester Institute of Technology, 2009.
43. Sumanasena, MG Buddika. A multidimensional systems approach to grid sensor networks. University of Notre Dame, 2012.
44. Carter, Brian. Multicriteria large scale heterogeneous sensor network deployment framework. University of Louisville, 2009.

References

1. D.W. Heuven, Opportunistic sensing and aggregation network using smartphones and sensor nodes. MS Thesis. University of Twente, 2014
2. U. Naik, Implementation of Distributed Composition Service for Self-Organizing Sensor Networks. Dissertation, 2005
3. D. Jourdan, Wireless Sensor Network Planning with Application to UWB Localization in GPS-Denied Environments. Dissertation, Massachusetts Institute of Technology, 2006
4. R.S. Ivanov, Context-aware Sensor Fusion for Securing Cyber-Physical Systems. Dissertation, University of Pennsylvania, 2017
5. V. Subramanian, Building Distributed Data Models in a Performance-Optimized, Goal-Oriented Optimization Framework for Cyber-Physical Systems. Dissertation, State University of New York at Stony Brook, 2012
6. R.R. Brooks, S.S Iyengar, *Multi-Sensor Fusion: Fundamentals and Applications with Software* (Prentice-Hall, Englewood Cliffs, 1998)
7. A. Pathak, Compilation of Data-Driven Macroprograms for a Class of Networked Sensing Applications. University of Southern California, 2008
8. M.U. Ilyas, Analytical and Quantitative Characterization of Wireless Sensor Networks. Michigan State University, 2009

Part III
Trends of Brooks–Iyengar Algorithm

Chapter 6
Robust Fault Tolerant Rail Door State Monitoring Systems

6.1 Introduction

The Brooks–Iyengar hybrid algorithm [1] for distributed control in the presence of noisy data combines Byzantine agreement with sensor fusion. It bridges the gap between sensor fusion and Byzantine fault tolerance [2]. The algorithm is fault tolerant and distributed. It could also be used as a sensor fusion method. The precision and accuracy bound of this algorithm have been proved in 2016 [3]. This seminal algorithm unified these disparate fields for the first time. Essentially, it combines Dolev's algorithm for approximate agreement [4–6] with Mahaney and Schneider's fast convergence algorithm (FCA) [7]. Researchers have also extended the original Byzantine agreement to Byzantine Vector Consensus (BVC) [8, 9]. The algorithm assumes N processing elements (PEs), τ of which are faulty and can behave maliciously. It takes as input either real values with inherent inaccuracy or noise (which can be unknown), or a real value with a priori defined uncertainty, or an interval. The output of the algorithm is a real value with an explicitly specified accuracy. The algorithm runs in $O(NlogN)$ where N is the number of PEs. It is possible to modify this algorithm to correspond to Crusader's Convergence Algorithm (CCA) [7]; however, the bandwidth requirement will also increase. There are several emerging problems and challenges in in the context of critical infrastructure operation [10], energy systems [11], complex networks [12], IoT applciations [13], and infrastructure resilience [14] that require efficient computational techniques, especially efficient distributed sensing. The algorithm has applications in distributed control, software reliability, and high-performance

The following article with permission has been reproduced from the original copy: Ao, Buke. "Robust Fault Tolerant Rail Door State Monitoring Systems: Applying the Brooks–Iyengar Sensing Algorithm to Transportation Applications." International Journal of Next-Generation Computing 8.2 (2017): 108–114.

computing [7], and can be used to find the "fused" measurement of the weighted average of the midpoints of regions [15].

The Brooks–Iyengar algorithm has been used in a variety of redundancy applications [3, 16], including a program demonstration through the US Defense Advanced Research Projects Agency (DARPA) with BBN using the Sensor Information Technology for the War Fighter (SensIT) program. SensIT program develops software for networks of distributed microsensors, specifically in collaborative signal and information processing and fusion of data. More specifically, information was received from sensors in reconnaissance, surveillance, tracking, and targeting for battlefield operations. This work was an essential precursor to the Emergent Sensor Plexus MURI from Penn State University's Applied Research Laboratory (PSU/ARL), which incorporated the Brooks–Iyengar algorithm to extend SensIT's advances to create practical and survivable sensor network applications.

The Brooks–Iyengar algorithm has also been extended into modern-day LINUX and Android operating systems. In these applications, the algorithm combines data to provide fault tolerant data fusion which is used by 99% of the world's top supercomputers, 79% of all smartphones worldwide, and 100% of users accessing the Internet, to provide seamless operations and service.

6.2 Safety-Critical Transportation Applications

In this section of the chapter, we present the theoretical applications of fusion, then demonstrate a detailed application and simulation of the Brooks–Iyengar algorithm in solving the safety challenges of closing and opening the doors aboard moving trains. Accurately detect the state of train door such as the variable to be measured is critical to the application.

6.3 Theory

In a fusion system, we want to fuse different interval values from sensors where τ represents the number of them that are faulty. Suppose we have N sensors, which measures the variable value of $[l_1, h_1], \ldots, [l_N, h_N]$.

Notations:

- v_T: The ground truth value
- v: The output value of Brooks–Iyengar algorithm
- I_{BY}: The output interval of the Brooks–Iyengar algorithm
- $a_{N-2\tau}$: The left endpoint of the region where $N - 2\tau$ non-faulty intervals overlap
- $b_{N-2\tau}$: The right endpoint of the region where $N - 2\tau$ non-faulty intervals overlap
- g: A set of $N - \tau$ valid measurements
- f: A set of τ faulty measurements

- G: The set of all possible valid measurements, so $g \in G$
- F: The set of all possible faulty measurements, so $f \in F$

We build three theorems to describe the performance of Brooks–Iyengar algorithm in a fusion system. Theorems 6.1 and 6.2 identify the accuracy feature of Brooks–Iyengar algorithms. Theorem 6.3 is the comparison of Brooks–Iyengar algorithm with other related algorithms.

Theorem 6.1 *In a system of N sensors in which O of them are faulty, $v_T \in I_{BY} \subseteq [a_{N-2\tau}, b_{N-2\tau}]$ and*

$$|v - v_T| \le |I_{BY}| \le \min_{\tau+1}\{|v| : v \in g\}$$

where, $|I_{BY}|$ is the length of the output interval, v and I_{BY} is the output value and interval of Brooks–Iyengar algorithm.

Proof We assume all non-faulty intervals intersect on a common region, and we define it as I_{opt}. Since there are at least $N - \tau$ non-faulty intervals, so the weight of I_{opt} is at least $N - \tau$, and we have $I_{opt} \subseteq I_{BY}$, so $v_T \in I_{opt} \subseteq I_{BY}$. Since there are at most τ faulty intervals and the threshold of Brooks–Iyengar algorithm is $N-\tau$, so only regions that are formed by non-faulty intervals with weight equal or larger than $N - 2\tau$ intersect with τ faulty intervals could be subset of I_{BY}. So $[a_{N-2\tau}, b_{N-2\tau}]$ is the upper bound of I_{BY}, and $I_{BY} \subseteq [a_{N-2\tau}, b_{N-2\tau}]$. The equation, $|I_{BY}| \le \min_{\tau+1}\{|v| : v \in g\}$ has already been proved by Theorem 6.2 [17] as outlined by Marzullo, and as illustrated below.

Theorem 6.2 *The interval $[a_{N-\tau}, b_{N-\tau}]$ is smallest interval that is guaranteed to contain the true value [17].*

Proof The proof is in Algorithm 1 [17], as follows.

Therefore, the algorithm below provides the fusion value v of the Brooks–Iyengar algorithm, given the set of all possible valid measurements g and all possible faulty measurements f:

$v = BY(g, f)$: The fusion value v of Brooks–Iyengar algorithm given g and f

Theorem 6.3 *In a system of N sensors in which τ of them are faulty, then we have two sets G and F, $v = BY(g, f)$, where $g \in G$ and $f \in F$, then we have*

$$\max_{g,f} |v - v_T| \le \max_{g,f} |v_{ABA} - v_T|$$

$$\max_{g,f} |v - v_T| \le \max_{g,f} |v_{BVC} - v_T|$$

$$\max_{g,f} |v - v_T| \le \max_{g,f} |v_{avg} - v_T|$$

where v_{ABA}, v_{BVC}, and v_{avg} are the results of approximate Byzantine agreement (ABA), Byzantine Vector Consensus (BVC) [8], and naive average algorithm, here we calculate the midpoints of the intervals as inputs of the two algorithms.

Proof From Theorem 6.1 we know that, $v_T \in I_{BY} \subseteq [a_{N-2\tau}, b_{N-2\tau}]$, and from Proposition 4.1 in [3], we know there exists $g \in G$ and $f \in F$ such that $v_{ABA} \notin [a_{N-2\tau}, b_{N-2\tau}]$, which means $\max_{g,f} |v_{ABA} - v_T| \geq \max(|b_{N-2\tau} - v_T|, |a_{N-2\tau} - v_T|) \geq \max_{g,f} |v - v_T|$. So we have $\max_{g,f} |v - v_T| \leq \max_{g,f} |v_{ABA} - v_T|$. Similarly, from Proposition 4.2 in [3], there exists g, f such that $v_{BVC} \notin [a_{N-2O}, b_{N-2O}]$, and we could easily prove $\max_{g,f} |v - v_T| \leq \max_{g,f} |v_{BVC} - v_T|$. Equation (2) of [3] shows that the naive average could not tolerant fault so $\max_{g,f} |v_{avg} - v_T| = \infty$ and then $\max_{g,f} |v - v_T| \leq \max_{g,f} |v_{avg} - v_T|$.

6.4   Implementation

In this section, we show a cluster of sensors being used to detect the state of opening and closing of the train's doors. The situation could involve more than one hundred sensors, where as many as one third of the sensors could be faulty, yet using the Brooks–Iyengar Algorithm, we can still maintain accurate results.

Accurately detecting the state of the train's door is very crucial for ensuring people's safety. However, traditional scheme uses a single sensor to detect the current or some other variables of the train's door system which is not accurate when the sensor behaves fault. By leveraging Brooks–Iyengar algorithm, we could use multiple sensors to measure variables (current, etc.) robustly and accurately.

For example, let us use four sensors, where one of them is faulty. Each sensor's output is $[v_i - \sigma, v_i + \sigma]$, $1 \leq i \leq 4$, where the uncertainty bound for non-faulty and faulty sensors is identical for simplicity, so the ground truth v_T could be any point within the bound for non-faulty sensors but the output of fault sensor may not contain the ground truth v_T. The simulation is as follows.

Figure 6.1 gives an example of the output of Brooks–Iyengar algorithm. The green line is ground truth that has been pre-defined. We have four sensors to measure the variable, where one of them is faulty. The blue line is the fused value of Brooks–Iyengar Algorithm and we could also find that the ground truth lies in the upper bound and lower bound of Brooks–Iyengar algorithm.

Figure 6.2 shows a comparison between Brooks–Iyengar algorithm and the naive average algorithm. We assume that the distance between ground truth and faulty sensor reading is Gaussian distribution. Then we run two algorithms in the same condition. From the simulation results, we could find that the bound of Brooks–Iyengar algorithm always contains the ground truth, while the output of naive average sometimes is far from the ground truth. Since the bound of Brooks–Iyengar algorithm smallest bound to contain the ground truth, the green line that is not in the bound must be faulty outputs, which are denoted by the red points.

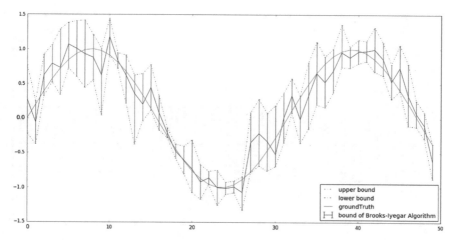

Fig. 6.1 The output of Brooks–Iyengar algorithm

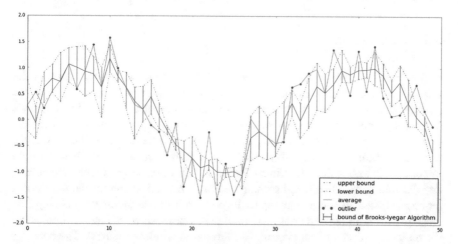

Fig. 6.2 Comparison of Brooks–Iyengar algorithm and average

Figure 6.3 considers the output of Brooks–Iyengar algorithm and naive average, when the faulty sensor reading keeps increasing. We could find that when the faulty sensor readings increase, the output of naive average algorithm becomes larger and larger, which indicates its sensitive to bad sensor readings. Both the output value and bound of Brooks–Iyengar algorithm are robust when the faulty sensor reading becomes larger, which shows that the Brooks–Iyengar algorithm is robust to outlier or faulty sensor readings.

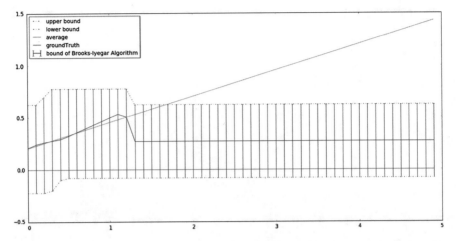

Fig. 6.3 The output of two algorithms when faulty values increase

6.5   Conclusion

In this train example, with acceleration and deceleration adversely affecting the sensor systems, the authors induced an error in one of the sensors to examine the effectiveness of the Brooks–Iyengar Algorithm in these applications. Since today's technology does not guarantee success and safety in all situations, the Brooks–Iyengar algorithm can significantly improve the fault tolerance of these systems, providing a greater margin of safety for operations. In doing so, there are two important situations that are critical for the safety of passengers when embarking and disembarking. Distributed sensing networks needed to control the train doors require a fusion of the sensor inputs to provide accurate automatic opening and closing with minimum traction. An automatic sensor network can be installed in the motor circuit to collect current data through a wireless protocol. The data can be transmitted by cellular communication to servers, where the Brooks–Iyengar distributed sensing algorithm can be applied to identify and categorize the data signals to safely and automatically open and close the doors.

In this chapter, the authors described and demonstrated the performance evaluation of the signal output of Brooks–Iyengar algorithm in this application. Based upon the performance results, the Brooks–Iyengar Algorithm provides the best robust algorithm for implementation under faulty sensor conditions, such as those encountered in real-world transportation applications.

References

1. R.R. Brooks, S.S. Iyengar, Robust distributed computing and sensing algorithm. Computer **29**(6), 53–60 (1996)
2. M. Ilyas, I. Mahgoub, L. Kelly, *Handbook of Sensor Networks: Compact Wireless and Wired Sensing Systems* (CRC Press, Boca Raton, 2004)
3. B. Ao, Y. Wang, L. Yu, R.R. Brooks, S.S. Iyengar, On precision bound of distributed fault-tolerant sensor fusion algorithms. ACM Comput. Surv. **49**(1), 5:1–5:23 (2016)
4. D. Dolev, The Byzantine generals strike again. Technical report, Stanford, CA, USA, 1981
5. L. Lamport, R. Shostak, M. Pease, The Byzantine generals problem. ACM Trans. Program. Lang. Syst. **4**(3), 382–401 (1982)
6. D. Dolev, N.A. Lynch, S.S. Pinter, E.W. Stark, W.E. Weihl, Reaching approximate agreement in the presence of faults. J. ACM **33**(3), 499–516 (1986)
7. S.R. Mahaney, F.B. Schneider, Inexact agreement: accuracy, precision, and graceful degradation, in *Proceedings of the Fourth Annual ACM Symposium on Principles of Distributed Computing, PODC '85* (ACM, New York, 1985), pp. 237–249
8. N.H. Vaidya, V.K. Garg, Byzantine vector consensus in complete graphs, in *Proceedings of the 2013 ACM Symposium on Principles of Distributed Computing, PODC '13* (ACM, New York, 2013), pp. 65–73
9. H. Mendes, M. Herlihy, Multidimensional approximate agreement in Byzantine asynchronous systems, in *Proceedings of the Forty-fifth Annual ACM Symposium on Theory of Computing, STOC '13* (ACM, New York, 2013), pp. 391–400
10. M.H. Amini, Distributed computational methods for control and optimization of power distribution networks, Ph.D. Dissertation, Carnegie Mellon University, 2019
11. M.H. Amini, et al, Load management using multi-agent systems in smart distribution network, in *IEEE Power and Energy Society General Meeting* (IEEE, Piscataway, 2013), pp. 1–5
12. M.H. Amini (ed.), in *Optimization, Learning, and Control for Interdependent Complex Networks*. Advances in Intelligent Systems and Computing, vol. 2 (Springer, Cham, 2020)
13. A. Imteaj, M.H. Amini, Distributed sensing using smart end-user devices: pathway to federated learning for autonomous IoT, in *Proceeding of 2019 International Conference on Computational Science and Computational Intelligence*, Las Vegas (2019)
14. A. Imteaj, M.H. Amini, J. Mohammadi. Leveraging decentralized artificial intelligence to enhance resilience of energy networks (2019). arXiv preprint:1911.07690
15. S. Sahni, X. Xu, Algorithms for wireless sensor networks. IJDSN **1**, 35–56 (2005)
16. V. Kumar, Computational and compressed sensing optimizations for information processing in sensor network. IJNGC **3**, 1–5 (2012)
17. K. Marzullo, Tolerating failures of continuous-valued sensors. ACM Trans. Comput. Syst. **8**(4), 284–304 (1990)

Part IV
Applications of Brooks–Iyengar Algorithm for the Next 10 Years

Chapter 7
Decentralization of Data-source Using Blockchain-Based Brooks–Iyengar Fusion

7.1 Introduction

Information fusion when working with wireless sensor networks plays an important role. Sensors have evolved over time [1]. WSNs are usually deployed in large numbers and in various applications and infrastructures (e.g., smart cities and energy systems [2–4]) where they are exposed to highly varying conditions making their sensed values sometimes imprecise. Even in ambient conditions, the sensed values could be far from precise due to the possible failures. Such inoperable nodes can be possible nodes for attack and subsequent compromise. A more common issue with using WSN is the limited spatial and temporal coverage they can have in the environment. To overcome the above-mentioned limitations, previous works have classified the information fusion into co-operative, complementary, and redundant based on the relationship they have with the information source.

When information from multiple sources have to be aggregated to form the broader scene, the complementary algorithm is used. In cases when two or more sensor sources provide information about the same environment, we use the redundant algorithms. This provides the means for fault tolerance and pieces of the input source can be fused to increase associated confidence. Complementary fusion algorithms search for completeness by compounding new information from different pieces. This is highly applicable in a WSN-like environment. The most popular sensor fusion algorithms include *Kalman filter* (widely used in signal processing), *Bayesian networks* (used in Neural networks), *Dempster–Shafer* (probability calculus), and *Brooks–Iyengar algorithm*.

Major part of this chapter is reprinted from the following article: Iyengar, Sitharama S., Sanjeev Kaushik Ramani, and Buke Ao. "Fusion of the Brooks–Iyengar Algorithm and Blockchain in Decentralization of the Data-Source." Journal of Sensor and Actuator Networks 8.1 (2019): 17. Link to license: https://creativecommons.org/licenses/by/4.0/. Link to the article: https://doi.org/10.3390/jsan8010017.

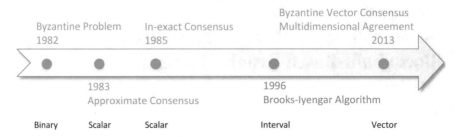

Fig. 7.1 Related algorithms (adopted from Wikipedia)

The Brooks–Iyengar algorithm being a distributed algorithm improves both the precision and accuracy of the sensor readings from a distributed WSN. This algorithm works well even in the presence of faulty inputs from sensors. This fault tolerance is achieved by exchanging the measured value and accuracy value at every node with every other node. These values are aggregated to calculate the accuracy range and a measured value for the whole network from all of the individual values collected. In 2016, the precision and accuracy bounds of this algorithm were proved. Because of its distributed and fault tolerant nature, it finds applications where the blockchains can be utilized. Various applications and dissertations based on the Brooks–Iyengar algorithm have proved this.

Figure 7.1 depicts the timeline of the various related algorithms. The list is as follows:

- *1982*—The Byzantine general Problem [5] was defined as an extension of the Two-Generals problem so that it can be viewed as a binary problem.
- *1983*—Approximate Consensus method [6] was introduced that eliminated values from the set of scalars to tolerate faulty inputs.
- *1985*—Inexact Consensus was introduced [7] that used scalar inputs.
- *1996*—Brooks–Iyengar algorithm [8] is an interval based approach, is highly fault tolerant and has since then seen many applications even until the date.
- *2013*—Byzantine Vector Consensus [9] as a method that works on vector inputs was introduced. The multidimensional agreement method was also introduced with the added feature that included the usage of distance.

Blockchains are distributed, decentralized public ledgers that deals with the storage and subsequent manipulation of digital information (blocks) in public databases (chains). One of the key features of a blockchain system can be explained in the way it is used in a Bitcoin like crypto-currency with the solution it provides to the double-spending problem based on the consensus algorithms and confirmation mechanisms. Blockchain systems and the corresponding network protocols have great potential in many fields including financial technology (FinTech), Internet of Things (IoT), smart grid, etc.

The blockchain system is a decentralized system, where every node has a complete copy of the blockchain data. This makes it more robust to malicious attacks and attempts to crack the system and extract information. Successful working without a centralized node which manages and maintains all the interactions showcases the lack of need for a specified official node to govern the database within the blockchain. All of the historical transactions among users are recorded as blocks and stored immutably in the blockchain system [10].

A typical blockchain system has the following features:

1. Every node has a full copy of the transactional records related to the application it is being used for right from the genesis of the application.
2. Participants are rewarded with a certain transaction fee or an incentive for any publicly verifiable computational work they perform. This forms the basis for the consensus, which acts as the backbone of the blockchain technology.
3. Repeated verification of the works of others enhances the chances of gaining the benefits or rewards and thus lures new nodes to join the network of decentralized nodes. The larger the number of nodes, the more are the number of people verifying the transactions and the better is the integrity of the data.

In the blockchain system, the transaction history is distributed to the decentralized peer-to-peer network. Each block has the details of transactions that have occurred. Hence, for a new block to be added to the parent chain and thus the blockchain system, it would need the consent of the participants in the chain. The new transaction that is reported is then packaged into a new block and awaits the successful acceptance by the participating nodes. After successful verification of the transaction recorded in the block, the block is added to the parent chain and every node updates its database with details of the new block.

7.1.1 Consensus in Blockchain

Consensus algorithms play an important role in the blockchain technology. It ensures that the integrity of the record in a block in the chain is maintained and the transaction is not tampered with or altered. In order to alter the transactional history in a block of an existing parent blockchain, the subchain would have to be completely altered and is a very tedious task to be performed in the midst of the many peering eyes of the fellow nodes. Thus, the historical records are consistent in every node.

Many consensus algorithms have been introduced to support the Blockchain system. Some of the well-known consensus algorithms are:

- Proof-of-Work (PoW);
- Proof-of-Stake (PoS);
- Delegated-Proof-of-Stake (DPoS);
- Practical Byzantine Fault Tolerance (PBFT), etc.

Each of these algorithms has its own set of advantages and disadvantages. The applicability of any of the above consensus algorithms depends on the application it is being used for. Thus, identifying the appropriate consensus algorithm that would fit the application is an important aspect of designing the blockchain system.

In this chapter, we use references to multiple sensors and sensor data and provide a novel technique of using the Brooks–Iyengar algorithm to decentralize the data source, which is the value in the transaction of a block, in case the data source is dominated by one or few groups. The rest of the chapter is organized as follows. Section 7.2 gives a brief introduction to blockchains and the various structures that exist in blockchains. Principles related to the blockchain technology are discussed in Sect. 7.3. The succeeding sections highlight the applicability of the Brooks–Iyengar algorithm in combination with a multi-sensor environment while using the blockchain system. The proposed procedure using the algorithm would provide a means to decentralize the data source. Finally, we give a theoretical analysis of the usage of the Brooks–Iyengar algorithm in this context and conclude the paper.

This paper describes a novel technique to explain the decentralization of a data source using a blockchain based approach. Information fusion from multiple sensors is explained using the seminal work of the Brooks–Iyengar Algorithm. The use of this algorithm when working with the blockchain technology is the main contribution of this paper. The applicability of the design to a smart grid environment with possible seamless collaboration by a financial system energy system and the consumer is explained in this paper. A theoretical analysis of the performance of the algorithm has been explained in this paper and more experimentation is a part of the future research directions.

7.2   Blockchain Structures

Blocks in a blockchain system hold batches of valid transactions. These transactions are encoded by a Merkle tree into hash values. Each block includes the hash values of the previous block and this chain runs until the path to the very first block of the chain can be traced. Thus, every two blocks in the system are linked like a chain giving it the name *Blockchain*. Figure 7.2 showcases the links between the individual blocks and how it grows into the complete blockchain. This chain of blocks is expected to have a great impact on the way things work. A good example of the applicability of blockchains is the way it has revolutionized the working of crypto-currencies. Many other researchers have discussed the various applications of blockchains. Newer versions or these have been introduced and implemented successfully.

As depicted in Fig. 7.2, each block consists of a block header and a list of transactions. The block header has six parameters that include:

- Version;
- Hash value of the previous root;
- Hash of the Merkel root;
- Timestamps;
- Number of bits and
- Nonces.

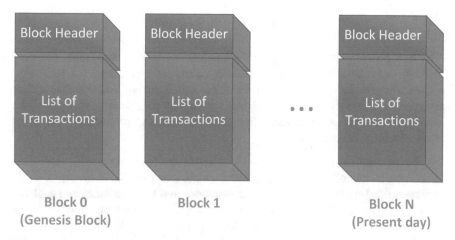

Fig. 7.2 Blockchain structures

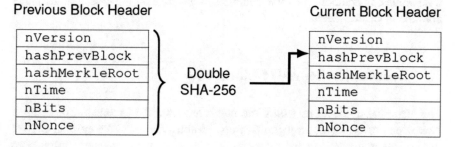

Fig. 7.3 Block header

The blocks are linked by the item of the *hashPrevBlock*. A detailed view of the block header with the various fields encompassed in it is depicted in Fig. 7.3, where the *hashPrevBlock* of $(n + 1)$th block is the Double SHA-256 hash value of the nth block. Finally, all blocks are linked and all historical transactions are recorded in the blockchain.

7.2.1 Transaction Procedure in Blockchain

The transaction procedure adopted in blockchains solved the double-spending problem by employing the consensus algorithm and replicating the confirmed stages. The process of finding a valid block is called mining, and is accomplished by specialized nodes called the mining nodes. The participants involved in the consensus process perform the mining and the process involved in this is as discussed in the following loop:

1. New transactions are generated and broadcast into the network;
2. Mining nodes collect these broadcasted transactions from the network;
3. The mining nodes then verify all transactions and package them into a block, making a record of all the inputs that have already been used previously;
4. Select the most recent block on the longest chain in the blockchain. This block would have the most votes from participant nodes. The hash value of the latest block determined in such a manner is inserted into the new block header;
5. *Solve the Proof-of-Work (PoW) problem*—If the PoW solution is checked indicating that all the transactions are verified, then the block is validated;
6. The new validated block is broadcasted to the network;
7. Each node that receives the new block will check the availability of the new block and verify the signature and PoW solution of theirs. If the block is checked and validated, the node will accept the new block and insert it into the blockchain where it becomes immutable and thus cannot be altered.

Every transaction to be added to the blockchain will need to be confirmed and checked by six or more nodes. This means that, if some malicious node wants to alter the transactions, it must reconfirm the new transactions and the corresponding hash values; in addition, the PoW solutions to generate all the blocks in the chain after the block are altered.

7.3 Transaction Source of Blockchain

The historical transactions within the blockchain and the transaction procedures have been proved robust through many applications. However, the most successful discussions in many years have been in the area of crypto-currencies. Introducing the blockchain architecture into new areas calls for more research and understanding of the system. Systems such as smart grids and Internet of Things (IoT), where the data produced or transmitted by the source of the transactions are not reliable, would need the use of other algorithms that could make the system more robust. For example, Fig. 7.4 shows the future energy system with the blockchain technology deployed in its functioning.

Figure 7.4 depicts a scenario wherein an energy supplier harnesses energy from multiple sources including nuclear, solar, wind, hydel, coal, and other methods. This

Concept – Future Energy System

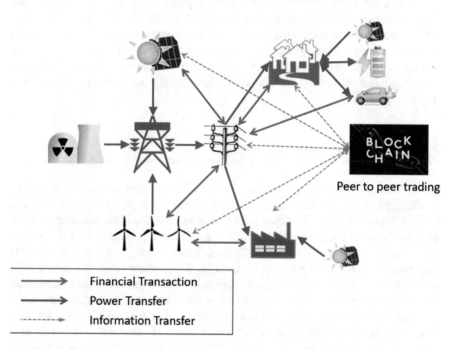

Peer to peer trading

→ Financial Transaction
→ Power Transfer
----→ Information Transfer

Fig. 7.4 Blockchain based smart grids [adopted from Department of Science and Technology grant—thanks to ProfGarg]

is the scenario with many suppliers who supply energy to a specific geographic area. With the advent of IoT and the benefits it has in collaboration with the smart grids, we can expect the energy delivering process to the household automated. In such cases, there would be the use of multiple sensors that monitor the amount of energy being supplied and it can be argued that each of the suppliers would have their own set of sensors to do this process.

In Fig. 7.4, the transactions need to be confirmed and consistent among all nodes such as users, financial institutes, and power stations. However, the source of the transactions which refers to the amount of the electricity consumed by the user is not completely decentralized and robust. As an example, the amount of electricity consumed is usually measured by a sensor. Any issue with the sensor or its malfunctioning, or possibility of attack by some adversaries and thus the alteration of the input raw data, makes the source unreliable.

In addition, in such scenarios, falsification of data as the sensor reading and thus reported to the blockchain system could lead to catastrophes. Therefore, we need a solution that can make the data source for the transaction fully decentralized and robust.

Thus, it is necessary to have a novel technique of decentralizing the data source and also make sure the process is robust and highly immune to faulty transactions. There is also a need for a strong fusion algorithm that would help in harnessing all the benefits that a distributed ledger like blockchains has to offer. The following section provides a brief introduction to the seminal work on the Brooks–Iyengar algorithm that would be used in designing a novel technique to decentralizing the data source and thus makes the blockchain system built for such applications more robust.

7.3.1 Brooks–Iyengar Algorithm

One of the seminal algorithms that has had a profound impact on sensor technology applications is the Brooks–Iyengar distributed sensing algorithm [8, 11, 12]. The Brooks–Iyengar hybrid algorithm for distributed control in the presence of noisy data combines Byzantine agreement with sensor fusion. It bridges the gap between sensor fusion and Byzantine fault tolerance. This seminal algorithm unified these disparate fields for the first time. Essentially, it combines Dolev's algorithm for approximate agreement with Mahaney and Schneider's fast convergence algorithm (FCA). The algorithm assumes N processing elements (PEs), t of which are faulty and can behave maliciously. It takes as input either real values with inherent inaccuracy or noise (which can be unknown), or a real value with a priori defined uncertainty, or an interval. The output of the algorithm is a real value with an explicitly specified accuracy. The algorithm runs in O (NlogN), where N is the number of PEs: see Big O notation. It is possible to modify this algorithm to correspond to Crusader's Convergence Algorithm (CCA); however, the bandwidth requirement will also increase. The algorithm has applications in distributed control, software reliability, and high-performance computing.

Initially, it was observed that mapping a group of sensor nodes to estimate its value accurately would need all the sensors to mutually exchange the values amongst themselves. This was a computationally expensive process. The algorithm utilized an array of sensors that sensed a similar environment and passed the collected data onto a virtual sensor node. This virtual node thus defined would collect data from all the sensors and aggregate the data from un-corrupted sensors. The Byzantine algorithm [5] and its fault tolerant methodology when used in such situations provides the necessary solution when working with real-time data from multi-sensor applications. This brought the rise of the Brooks–Iyengar algorithm or the Brooks–Iyengar hybrid algorithm that provides solutions to enhance the precision and accuracy of the measurements taken by a distributed sensor network. This algorithm works well even in the presence of faulty sensors [13].

The Brooks–Iyengar algorithm can fuse multiple sensors in a local system or reach consensus in distributed systems with the capability of fault tolerance. In case the source of data for the blockchain is unreliable, we would need the benefits of the Brooks–Iyengar algorithm to make the data source decentralized and robust to malicious attack or sensor errors.

7.3.2   Combining the Brooks–Iyengar Algorithm with Blockchains

Since blockchains have no official node, and the application deals with decentralized network of nodes, there is a need for a robust consensus algorithm. By using such a consensus based approach in the smart grid example, we can introduce multiple suppliers to the measurement stage of the consumption by a user. More specifically, multiple sensors would have to be employed to measure the physical world and make a *decentralized* measurement that will ensure that the values in the transactions of the blockchain is robust and the integrity is maintained even when distributed among all nodes in the network.

In similar lines, to visualize the application depicted in Fig. 7.4, we would use multiple heterogeneous sensors from different suppliers to measure the physical values in its surroundings. The Brooks–Iyengar algorithm would work in conjunction and attempt to fuse the measurements of the physical values. Redundant sensors from different suppliers measure the physical values according to their own criteria, and using the Brooks–Iyengar algorithm can make the sensor readings consistent. Figure 7.5 shows the use of four sensors to measure the electricity flow and supply. Each of these sensors is installed by varying suppliers, who, with their independent sensors, can avoid the data source from being determined by one or few predominant groups that try to monopolize the system and its activities.

We combine the Brooks–Iyengar algorithm with blockchain by adopting the following steps:

- Add multiple sensors that can individually collect raw sensor data.

  Each supply has an independent sensor or group of sensors to perform this action. Redundant and independent sensors enhance the fault tolerance ability and make the data source decentralized.

- Introducing the Proof of Work (PoW) consensus mechanism to provide each node with the permissions and ability to check and verify the raw sensor data.
- After careful verification and with the consent of the multiple participating nodes, the raw data from different suppliers are saved into blocks and ascertained to be a part of the parent chain of the Blockchain.

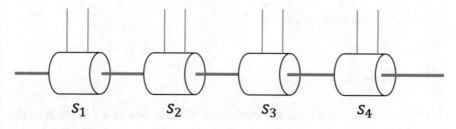

Fig. 7.5 Multiple current transformers to collect data

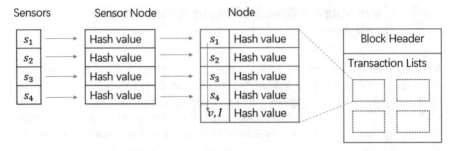

Fig. 7.6 Data from sensor to block example

The sensor node measures the physical values and computes a hash value of the raw data. These hash values are later uploaded to the blockchain. The virtual node collects the sensor data from multiple sensors and computes to find the solution (raw sensor data) of the hash values and then uses the Brooks–Iyengar algorithm to get the result. Finally, a transaction that includes the solution (raw data), hash values, and the result of the Brooks–Iyengar algorithm will be packaged into a block and broadcasted through the network.

The algorithm combines the benefits of the Brooks–Iyengar algorithm and blockchain technology as follows:

1. Each sensor node encodes the measurements into a block of hash values and uploads it.
2. Each node collects the packets from sensor nodes and tries to identify the real measurements, following which, the Brooks–Iyengar algorithm is used to fuse the multiple sensors' readings.
3. Real measurements, hash values, and the result of the Brooks–Iyengar algorithm are packaged into a block.
4. Each node receives the new block and checks and verifies the authenticity of the transaction and decides to add it to the parent or dominant chain of the blockchain or not.

Figure 7.6 shows an example of the packet from sensor to block with four sensors, where $s_1, \ldots, s_4$ are the raw measurements and v, I are the outputs of the Brooks–Iyengar algorithm. The complexity of the encoding could be adapted to the real scenario, as well as the raw data could also be chained or linked one-by-one to be exempted from cracking or forging.

7.3.3 Example

There are various advantages of using the blockchain technology in smart grids [14, 15]. In this section, we give an example of the Brooks–Iyengar algorithm to blockchain and verify its applicability in smart grids. The platform used including

time

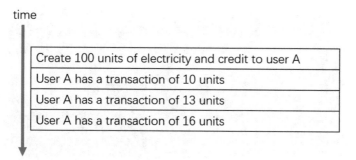

| Create 100 units of electricity and credit to user A |
| User A has a transaction of 10 units |
| User A has a transaction of 13 units |
| User A has a transaction of 16 units |

Fig. 7.7 Transactions over the timeline

software and hardware such as the smart metering module is open-source and decentralized, i.e., anyone can set a sensor to measure the electricity consumption. Each user has an account that displays the records corresponding to the electricity balance and electricity consumption each day. The energy consumption is uploaded, validated, and persisted in the blockchain where all the historical transactions are housed.

In a typical timeline of transactions, let us assume that the user has a balance of 100 units granted by the electricity plant. The user may consume a certain amount of electricity units and the transactions pertaining to this use should be synchronized and validated between the user and electricity supplier or plant. A sample of the timeline showing a series of transactions that depicts the events that occur based on the consumption is as shown in Fig. 7.7.

Since hardware and software of the sensors and the nodes are open-source, each measurement point may consist of multiple sensor metering nodes. The multiple sensor metering nodes may come from different manufacturers and vendors or the user could place his own sensor metering, making it highly heterogeneous. In Fig. 7.8, there are four sensors that may be placed by different companies to measure the electricity consumption. The sensor value after being collected is fused and stored as transactions in the parent chain of the blockchain.

As shown in Fig. 7.8, four sensors (assuming they are non-faulty) are used to measure the energy value and get the corresponding hash value. The hash value is then decoded by the nodes, which is a simplified Proof of Work (PoW) mechanism. After the hash value of the four sensors is decoded, the values are fused by the Brooks–Iyengar algorithm and a corresponding transaction is created. Following this, the transactions are broadcast into the whole network like a typical blockchain transaction. Finally, the transaction is stored in the blockchain and the user and the electricity plant will reach an agreement on the units consumed and remaining for the user.

Table 7.1 showcases an example of values from multiple sensors with one of them being faulty. A sensor senses and constantly measures the electricity consumption. We have selected a snapshot of the sensor readings as measured at time t_0. These

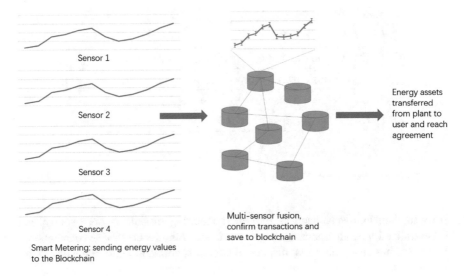

Fig. 7.8 From sensor to blockchain

Table 7.1 Sensors and the corresponding values

Sensor	Sensed value
S_1	[2.7, 6.7]
S_2	[0, 3.2]
S_3	[1.5, 4.5]
S_4	[0.8, 2.8]
S_5	[1.4, 4.6]

Table 7.2 Hashing the sensor values

Sensor	Hash of sensed value
S_1	f4053cf34bb2da93
S_2	10a11f0fb5ba6e81
S_3	3713b723d0308815
S_4	c695986ea057e573
S_5	8389c34490f813b0

are the readings that give an insight into the electricity consumption measurement from the user to the blockchain. At time t_0, we get values from five different sensors with s_1 being faulty.

After the sensor values have been obtained, we pass it through a known hashing algorithm to get the hash value of the sensor values and the data packet is broadcast randomly to some of the participating nodes. Table 7.2 depicts the hash values that have been created and the corresponding sensors that provided the input to the hashing algorithm. The hash values are then sent across to multiple participating nodes.

Table 7.3 Decode and fuse

Sensor	Sensed value	Hash of sensed value
S_1	[2.7, 6.7]	f4053cf34bb2da93
S_2	[0, 3.2]	10a11f0fb5ba6e81
S_3	[1.5, 4.5]	3713b723d0308815
S_4	[0.8, 2.8]	c695986ea057e573
S_5	[1.4, 4.6]	8389c34490f813b0

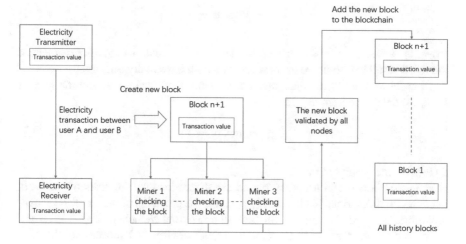

Fig. 7.9 Typical transaction process of a blockchain

The node receives the hash value, decodes the hash value, and tries to identify the real values. On identifying the initial value from the sensors, the Brooks–Iyengar algorithm is invoked to fuse the five sensors' values, and gets Table 7.3. In fact, s_1 is faulty and deviates from the fused value. s_r is the value obtained after the Brooks–Iyengar algorithm based data fusion is performed.

A transaction includes the data (transaction values) as shown in Table 7.3. Such transactions are broadcasted to other nodes and is validated by the other nodes like a typical transaction process in a blockchain.

Figure 7.9 shows the various steps involved in fulfilling the transaction process of a blockchain. The transaction values are the fused electricity consumption values measured by multiple sensors and combined according to the Brooks–Iyengar algorithm. The electricity transmitter is the user and the electricity receiver could be the electricity supply plant; a transaction has the details of the transaction made between them. The process includes creation of a new block, which is usually validated by a miner and broadcast to all other nodes. Finally, the block contains transaction values that are to be added to the longest surviving parent chain of the blockchain.

7.4   Accuracy and Precision of the Brooks–Iyengar Algorithm

As depicted in the previous example, the faulty sensors such as s_1 may have an impact and deviate the result, and we need to determine the precision and accuracy of the Brooks–Iyengar algorithm even in the presence of faulty sensors. We define the precision as

$$\varepsilon_{BY} = \max_{i,j} |v_i - v_j| \tag{7.1}$$

where v_i, v_j are the two sensor values. The precision could be the disagreement of the fused result between two users when they both have faulty sensors.

In order to identify the disagreement between the fused result and the ground truth, we define the accuracy as

$$\varsigma_{BY} = \max_{i} |v_i - \hat{v}| \tag{7.2}$$

where $\hat{v}$ is the ground truth.

Precision bound and accuracy bound are determined on the basis of the work published and described in [10]. Let ς_{BY} denote the precision of the Brooks–Iyengar algorithm, v_i, v_j are the value of ith and jth processing elements (PEs). The largest possible disagreement about the fusion output between two processing elements (PEs) is as depicted in the below equation

$$\varepsilon_{BY} = \max_{i,j} |v_i - v_j| \tag{7.3}$$

Accordingly, as per a well-known theorem from [10]

$$\max_{i,j} |v_i - v_j| = \frac{1}{1+\alpha} (b^g_{N-2\tau} - a^g_{N-2\tau}) \tag{7.4}$$

where $\alpha = \frac{N-\tau}{((2N-\tau)\tau)}$, $\tau \leq \frac{N}{3}$, N is the number of PEs, τ is the number of faulty PEs, and g is the set of non-faulty PEs.

Let ς_{BY} be the accuracy of the Brooks–Iyengar algorithm, $\hat{v}$ is the real value to be estimated, and then

$$\varsigma_{BY} = \max_{i} |v_i - \hat{v}| \leq \min_{\tau+1}\{|v| : v \in g\} \tag{7.5}$$

where $\min_{\tau+1}\{|v| : v \in g\}$ is the length of the $(\tau + 1)$th smallest measurement of all non-faulty sensors.

Accuracy and precision bound describes the property of the final value stored in the blockchain, and could be used further as reliable transactions.

7.4.1 Fault Tolerance

Since some sensors could be faulty and deviate from the fused result, we will show the fault tolerant ability of the Brooks–Iyengar algorithm when it is used as a sensor fusion method. The Brooks–Iyengar algorithm could tolerate $N/2$ faulty sensors and it is proved by Marzullo [16]; the output bound of the interval is

$$\text{If } \tau < \lfloor \frac{n+1}{2} \rfloor, \text{ then the output is bounded by } \min_{2\tau+1}\{|s| : s \in S\}$$

where τ is the faulty sensors number and n is the number of all sensors, and $\min_{2\tau+1}$ is the $(2\tau + 1)$th smallest values of all the input intervals.

7.5 Blockchain Architecture

A typical blockchain architecture contains many different layers, e.g., consensus layer, smart contract layer, communication layer, and data store abstraction. There are many different kinds of consensus algorithms, such as Proof-Of-Work (PoW), Proof-Of-Stake (PoS), Practical Byzantine Fault Tolerance (PBFT), delegated Proof-of-Stake (DPoS). Each of the consensus algorithm has unique feature and could be used in different systems, such as the PoW formed the Bitcoin system and the PBFT is the basic algorithm of Hyperledger project.

The consensus algorithm is an essential part in blockchain, which could assure the blockchain in different PE reach agreement. Brooks–Iyengar algorithm as an extension of Byzantine agreement could also be used in a blockchain system. Although most of the current blockchain system is based on exact agreement, Brooks–Iyengar algorithm that could reach approximate agreement could enable a new type of blockchain system. For example, if the block is formed by signal or sensor data, then we could use the Brooks–Iyengar algorithm to make sure the agreement in a distributed network.

Time synchronization is another issue in blockchain system, while current blockchain timestamp is usually comes from the network time protocol (NTP). This kind of clock source is easy while is not flexible in the blockchain application system. The foregoing chapters show that Brooks–Iyengar algorithm is also a robust time synchronization algorithm. So we could use it as the time synchronization algorithm of any blockchain system, which contributes to the fault tolerance and adapt to different scenario in the aspect of network error.

7.6   Modeling of Blockchain Assignment Based on Byzantine Consensus

The system is made up of a set P of n asynchronous sequential processes $\{p.\}$ which proceed at their own computational power and load and may be different with one another. The local processing time of each process is considered to be zero as it is much smaller than the delay of the network exchanging messages between the processes.

The processes communicate by exchanging messages through an asynchronous reliable point-to-point network. We assume that the network is *asynchronous*; i.e., the network delay is not infinite; but has no predefined maximum. Also, it is *reliable* in the sense that that it guarantees the delivery of messages as they are. Moreover, the connections between any pair of nodes in the network are bidirectional.

7.6.1   Byzantine Consensus Problem Based on Monte Carlo

In the Monte Carlo consensus algorithm, the agreement of two processes to decide the same block is randomized. In other words, no two connected processes will choose different blocks with the chance of at-least ϵ which is the high-threshold for the probability of agreement on a block for every pair of communicating processes.

One way of implementing the Monte Carlo Consensus Algorithm is to restrict the processes $p \in P$ and $q \in P$ to decide on the set of non-Byzantine blocks $R = \{r_i\}_1^k$ such that every r_i is a Byzantine block with the probability of $\frac{\tau}{n}$ and processes p and q decided r_i with the probability x_i.

The probability that both p and q select the same block in a naive Monte Carlo consensus algorithm would be

$$Pr[r_1 \text{ is chosen} \vee r_2 \text{ is chosen} \vee \ldots \vee r_k \text{ is chosen}] = \sum_{i=1}^{k} Pr[r_i \text{ is chosen}]$$

$$= \sum_{i=1}^{k} Pr[p \text{ chooses } r_i \wedge q \text{ chooses } r_i]$$

$$= \sum_{i=1}^{k} x_i^2$$

Therefore, the low-threshold of agreement probability will be

$$\epsilon \leq \sum_{i=1}^{k} x_i^2$$

7.7   A Fuzzy Byzantine Consensus

In this section, we propose a novel fuzzy Byzantine consensus for blockchain using the Brooks–Iyengar fusion algorithm. We prove how this approach handles fault tolerance in the presence of Byzantine processes.

Let $b_0, b_1, \ldots, b_k$ denote the chain of blocks of consecutive instances $0, 1, \ldots, k$ such that block b_{i+1} has the hashed content of block b_i for every $i = 0, 1, \ldots, k-1$ and b_0 represents the genesis block. Also, assume that processes $p_1, p_2, \ldots, p_n$ create a distributed system that uses the blockchain as a distributed database. We assume that every process can be a Byzantine with the probability of μ.

Here, we propose a novel solution based on Brooks–Iyengar fusion algorithm that make the processes to reach consensus on whether or not to agree on appending a new block $\mathcal{B}$ of transactions. The assumption is that every process has a copy of $\mathcal{B}$ and the hashing function verifying its correctness.

Assuming that every process p_i honestly votes Yes or No to the proposal of appending $\mathcal{B}$, we use the notation $v_i(\mathcal{B})$ to denote the indicator of the honest vote made by process p_i. Based on the assumption of Byzantine processes, the public vote of p_i on $\mathbb{B}$ is represented by $\hat{v}_i(\mathcal{B})$. Also, assume that every Byzantine process colludes with the probability of κ. Therefore

$$\mathbf{Pr}\left[\hat{v}_i(\mathcal{B}) = v_i(\mathcal{B})\right] = 1 - \kappa\mu$$

Utilizing Brooks–Iyengar fusion function f_{BI} for integrating the votes of processes $p_1, p_2, \ldots, p_n$, the *fuzzy consensus* would be

$$f_{\mathrm{BI}}\left(\hat{v}_1(\mathcal{B}), \hat{v}_2(\mathcal{B}), \ldots, \hat{v}_n(\mathcal{B})\right)$$

Theorem 7.1 *The fuzzy consensus of processes $p_1, p_2, \ldots, p_n$ utilizing Brooks–Iyengar fusion algorithm follows the normal distribution mentioned in (7.1) assuming that n is a large integer, γ is the ratio of block appending proposals that are cryptographically valid, and $\overline{\gamma} \ll 1$ ($\overline{x} = 1 - x$ for every real number x)*

$$f_{BI}\left(\hat{v}_1(\mathcal{B}), \hat{v}_2(\mathcal{B}), \ldots, \hat{v}_n(\mathcal{B})\right) \sim \mathcal{N}(\gamma + \kappa\mu\overline{\gamma}, \overline{\kappa\mu} \cdot \overline{\gamma}) \tag{7.6}$$

Proof The expected value of such fuzzy consensus does not depend on the normalized weights of fusion function f_{BI} since $\sum_{i=1}^{n} w_i = 1$

$$\mathbf{E}\left[f_{\mathrm{BI}}\left(\hat{v}_1(\mathcal{B}), \hat{v}_2(\mathcal{B}), \ldots, \hat{v}_n(\mathcal{B})\right)\right] = \pi \sum_{i=1}^{n} w_i = \pi$$

where π is the probability of Yes vote of a process that can be computed in the following way

$$\pi = \mathbf{Pr}\Big[(\text{real vote is Yes}) \vee (\text{real vote is No and the process is dishonest})\Big]$$

$$= \gamma + \kappa\mu\bar{\gamma}$$

Also, the standard deviation of the consensus is not a function of weights

$$\mathbf{STD}\Big[f_{\mathbf{BI}}\big(\hat{v}_1(\mathcal{B}), \hat{v}_2(\mathcal{B}), \ldots, \hat{v}_n(\mathcal{B})\big)\Big] = \sqrt{\sum_{i=1}^{n} w_i \pi \bar{\pi}}$$

$$\approx \begin{cases} \sqrt{\gamma + (1 - 3\gamma)\kappa\mu} & \text{if } \gamma \ll 1; \text{ i.e. most proposals are not valid;} \\ \sqrt{\kappa\mu \cdot \bar{\gamma}} & \text{if } \bar{\gamma} \ll 1; \text{ i.e. most proposals are valid;} \\ \sqrt{\kappa\mu + (1 - 3\kappa\mu)\gamma - \gamma^2} & \text{otherwise} \end{cases}$$

Assuming that the number of processes n is large enough ($n \gg 1$) and most proposals are valid ($\bar{\gamma} \ll 1$, the probability distribution of fuzzy consensus can be estimated by normal distribution $\mathcal{N}(\gamma + \kappa\mu\bar{\gamma}, \overline{\kappa\mu \cdot \bar{\gamma}})$.   □

Theorem 7.2 *Assuming that μ portion of processes $p_1, p_2, \ldots, p_n$ are Byzantine, the fuzzy consensus of these processes utilizing Brooks–Iyengar fusion algorithm guarantees the following precision and recall ratio*

$$\text{False Negative} \leq 0.5\Big(1 - \text{erf}\Big(\frac{-0.5\mu}{\sqrt{\mu}}\Big)\Big) \tag{7.7}$$

$$\text{False Positive} \leq 0.5\Big(1 - \text{erf}\Big(\frac{-0.5}{\sqrt{\mu}}\Big)\Big) \tag{7.8}$$

Proof First, we calculate the true positive ratio

$$\text{recall} = \mathbf{Pr}\Big[f_{\mathbf{BI}}\big(\hat{v}_1(\mathcal{B}), \hat{v}_2(\mathcal{B}), \ldots, \hat{v}_n(\mathcal{B})\big) \geq 0.5 | \text{vote is Yes}\Big]$$

$$= \mathbf{Pr}\Big[f_{\mathbf{BI}}\big(\hat{v}_1(\mathcal{B}), \hat{v}_2(\mathcal{B}), \ldots, \hat{v}_n(\mathcal{B})\big) \geq 0.5 | \gamma \geq 0.5\Big]$$

$$\geq 0.5\Big(1 - \text{erf}\Big(\frac{0.5 - \gamma - \mu\bar{\gamma}}{\sqrt{2\mu \cdot \bar{\gamma}}}\Big)\Big) \text{ if } \gamma \geq 0.5$$

$$\geq 0.5\Big(1 - \text{erf}\Big(\frac{-0.5\mu}{\sqrt{\mu}}\Big)\Big)$$

Then, we calculate the false positive ratio

$$\text{FP} = \mathbf{Pr}\Big[f_{\text{BI}}\big(\hat{v}_1(\mathcal{B}), \hat{v}_2(\mathcal{B}), \dots, \hat{v}_n(\mathcal{B})\big) \geq 0.5 | \text{vote is No}\Big]$$

$$= \mathbf{Pr}\Big[f_{\text{BI}}\big(\hat{v}_1(\mathcal{B}), \hat{v}_2(\mathcal{B}), \dots, \hat{v}_n(\mathcal{B})\big) \geq 0.5 | \gamma \leq 0.5\Big]$$

$$\leq 0.5\Big(1 - \text{erf}\big(\frac{0.5 - \gamma - \mu\overline{\gamma}}{\sqrt{2\overline{\mu} \cdot \overline{\gamma}}}\big)\Big) \text{ if } \gamma \leq 0.5$$

$$\leq 0.5\Big(1 - \text{erf}\big(\frac{-0.5}{\sqrt{\overline{\mu}}}\big)\Big)$$

$\square$

7.8 Simulation

In this section, we will use an experiment [17], which uses four sensors to measure a physical signal, in order to explain the fault tolerant ability of the Brooks–Iyengar algorithm. Give a ground truth signal

$$y = \sin(\Omega t), 0 \leq t \leq 1 \tag{7.9}$$

where $\Omega = 1$ and the sampling interval is 0.2.

The sensors measure the signal and output an interval $[v_i - \sigma, v_i + \sigma], 1 \leq i \leq 4$, where v_i is the value and σ is the confidence range. Suppose $\bar{v}$ is the ground truth and three sensors' output values satisfy $|v_i - \bar{v}| \leq X$ and $X \sim U[0, 0.5]$, which means that the distance between ith measurement v_i and $\bar{v}$ is continuous uniform distribution $U[0, 0.5]$. One sensor is malfunction and $|v_f - \bar{v}| \sim U[0, 2.5]$, where v_f is the sensor's output value. Four sensors have the same quality and share the same confidence range $\sigma = 0.5$.

Given the above conditions, Fig. 7.10 shows a comparison between the Brooks–Iyengar algorithm and the naive average algorithm. Then, we run two algorithms in the same condition. From the simulation results, we find that the bound of the Brooks–Iyengar algorithm always contains the ground truth, while the output of naive average sometimes is far from the ground truth. Since the bound of the Brooks–Iyengar algorithm is the smallest bound to contain the ground truth, the green line that is not in the bound must be faulty outputs, which are denoted by the red points.

7.9 Conclusions

Consensus is a fundamental problem of distributed computing. Although this problem is proved to be NP-Hard, there have been many attempts to design protocols to either approximate consensus or solve it under various assumptions. However,

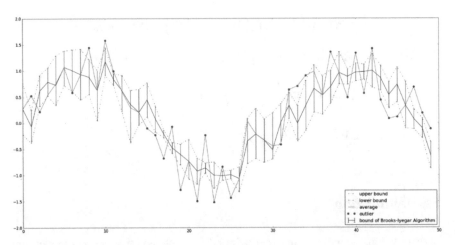

Fig. 7.10 Comparison of the Brooks–Iyengar algorithm and average

very little work has been devoted to explore its theoretical ramifications. Therefore, existing proposals are sometimes misunderstood and it is often unclear whether the problems arising during their executions are due to implementation bugs or more fundamental design issues. In this chapter, we proposed a fuzzy Byzantine consensus model for blockchains using the well-known Brooks–Iyengar fusion algorithm and theoretically showed how this approach handles faulty proposals of new blocks in the presence of Byzantine processes.

Also, the chapter describes the transaction protocol, consensus algorithm, and Proof of Work based consensus algorithm of the common blockchain. However, the common blockchain with only one sensor in applications like IoT or smart grids cannot decentralize the data source, thus making the transaction values in the blockchain controlled by a centralized node. In this chapter, we have introduced a novel technique that involves the Brooks–Iyengar algorithm and proposed a procedure of adapting it in a multi-sensor environment while decentralizing the data source. The paper also discusses the theoretical bound of the result of the Brooks–Iyengar algorithm, which is the value to be saved in the blockchain. Information fusion from multiple sensors is explained using the seminal work of the Brooks–Iyengar algorithm. The use of this algorithm when working with the blockchain technology is the main contribution of this paper. The applicability of the design to a smart grid environment with possible seamless collaboration by the financial system energy system and the consumer is explained in this paper. A theoretical analysis of the performance of the algorithm when used in a blockchain based decentralized environment has been explained in this paper and more experimentation is a part of the future research directions.

References

1. S.K. Ramani, S.S. Iyengar, Evolution of sensors leading to smart objects and security issues in IoT, in *International Symposium on Sensor Networks, Systems and Security* (Springer, Berlin 2017), pp. 125–136
2. M.H. Amini, Distributed computational methods for control and optimization of power distribution networks, PhD Dissertation, Carnegie Mellon University, 2019
3. M.H. Amini (ed.), in *Optimization, Learning, and Control for Interdependent Complex Networks*. Advances in Intelligent Systems and Computing, vol. 2 (Springer, Cham, 2020)
4. M.H. Amini, H. Arasteh, P. Siano, Sustainable smart cities through the lens of complex interdependent infrastructures: panorama and state-of-the-art, in *Sustainable Interdependent Networks*, vol. II. (Springer, Cham, 2019)
5. L. Lamport, R. Shostak, M. Pease, The byzantine generals problem. ACM Trans. Program. Lang. Syst. **4**, 382–401 (1982)
6. D. Dolev, N.A. Lynch, S.S. Pinter, E.W. Stark, W.E. Weihl, Reaching approximate agreement in the presence of faults. J. ACM **33**, 499–516 (1986)
7. S.R. Mahaney, F.B. Schneider, Inexact agreement: accuracy, precision, and graceful degradation, in *Proceedings of the Fourth Annual ACM Symposium on Principles of Distributed Computing*. Minaki (1985), pp. 237–249
8. R.R. Brooks, S. Iyengar, Robust distributed computing and sensing algorithm. Computer **29**, 53–60 1996
9. N.H. Vaidya, V.K. Garg, Byzantine vector consensus in complete graphs, in *Proceedings of the 2013 ACM Symposium on Principles of Distributed Computing*. Montréal (2013), pp. 65–73
10. B. Ao, Y. Wang, L. Yu, R.R. Brooks, S. Iyengar, On precision bound of distributed fault-tolerant sensor fusion algorithms. ACM Comput. Surv. **49**, 5 (2016)
11. S. Iyengar, R.R. Brooks, *Multi-Sensor Fusion: Fundamentals and Applications with Software* (Prentice-Hall, Englewood Cliffs, 1997)
12. B. Krishnamachari, S. vIyengar, Distributed Bayesian algorithms for fault-tolerant event region detection in wireless sensor networks. IEEE Trans. Comput. **53**, 241–250 (2004)
13. V. Kumar, Impact of Brooks–Iyengar distributed sensing algorithm on real time systems. IEEE Trans. Parallel Distrib. Syst. **25**, 1370–1370 (2014)
14. T. Winter, *The Advantages and Challenges of the Blockchain for Smart Grids*. Master thesis, Delft University of Technology, 2018
15. G.S. Thejas, T.C. Pramod, S.S. Iyengar, N.R. Sunitha, Intelligent access control: a self-adaptable trust-based access control (SATBAC) framework using game theory strategy, in *Proceedings of International Symposium on Sensor Networks, Systems and Security*, ed. by S.V.R Nageswara, R.R. Richard, Q.W. Chase (Springer, Cham, 2018), pp. 97–111. https://doi.org/10.1007/978-3-319-75683-7_7
16. K. Marzullo, Tolerating failures of continuous-valued sensors. ACM Trans. Comput. Syst. **8**, 284–304 (1990)
17. B. Ao, Robust fault tolerant rail door state monitoring systems: applying the Brooks–Iyengar sensing algorithm to transportation applications. Int. J. Next Gener. Comput. 8, 108–114 (2017)

Chapter 8
A Novel Fault Tolerant Random Forest Model Using Brooks–Iyengar Fusion

8.1 Introduction

Over the past two decades machine learning has become one of the mainstays of information technology and with that, a rather central, albeit usually hidden, part of our life. With the ever increasing amounts of data becoming available there is good reason to believe that smart data analysis will become even more pervasive as a necessary ingredient for technological progress.

8.2 Random Forest Classifiers

Definition 8.1 A random forest classifier is a classifier consisting of a collection of tree-structured classifiers $\{h_k(X, \Theta_k), k = 1, 2, \ldots\}$ where the $\{\Theta_k\}$ are independent identically distributed random vectors and each tree casts a unit vote for the most popular class at input x (See Fig. 8.1).

Considering a sequence of tree classifiers $h_1(x), h_2(x), \ldots, h_k(x)$ and assuming that random vector (Y, X) is the superset of randomly constructed training sets for the classifiers $h.$, we define the margin function $\mu(X, Y)$ in the following way:

$$\mu(X, Y) = \overline{\mathbf{I}[h_k(X) = Y]}_k - \max_{j \neq Y} \overline{\mathbf{I}[h_k(X) = j]}_k$$

where $\mathbf{I} : U \mapsto \{0, 1\}$ denotes the identifier function. The way that we define the marginal function implies that by as more trees (on average) predict the correct label Y than any other label, the margin function results in a higher value. In fact, $\mu(X, Y)$ is directly dependent on the level of confidence of the ensemble of understudied classifiers.

© Springer Nature Switzerland AG 2020
P. Sniatala et al., *Fundamentals of Brooks–Iyengar Distributed Sensing Algorithm*,
https://doi.org/10.1007/978-3-030-33132-0_8

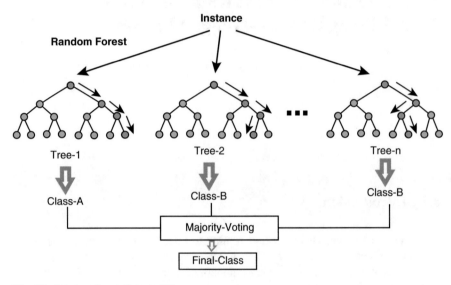

Fig. 8.1 Random forest diagram [1]

As long as the margin function $\mu(X, Y)$ is positive,

$$\mathbf{I}\big[h_k(X) = Y\big]_k > \max_{j \neq Y} \overline{\mathbf{I}\big[h_k(X) = j\big]_k}$$

which implies that the number of trees choosing the right label is more than those which have predicted any other label. Therefore, the ensemble of classifiers predicts a wrong class if and only if $\mu(X, Y) < 0$; i.e., the error value can be obtained by $\mathbf{Pr}[\mu(X, Y) < 0]$. Therefore,

$$\text{Error}(h_{1,\ldots,k}) = \mathbf{Pr}[\mu(X, Y) < 0] = \mathbf{Pr}\bigg[\mathbf{I}\big[h_k(X) = Y\big]_k < \max_{j \neq Y} \overline{\mathbf{I}\big[h_k(X) = j\big]_k}\bigg]$$

According to the strong law of large numbers, we have

$$\lim_{k \to +\infty} \text{Error}(h_{1,\ldots,k}) = \mathbf{Pr}\bigg[\mathbf{Pr}\big[h_k(X) = Y\big] < \max_{j \neq Y} \mathbf{Pr}\big[h_k(X) = j\big]\bigg] \qquad (8.1)$$

Equation (8.1) shows that even when the number of trees in a random forest classifier converges to infinity, there will be no overfitting as the error value does not converge to zero.

8.3 Enhanced Random Forest Regressors Utilizing Brooks–Iyengar Fusion Method

The idea of random forest classifiers is also applicable to regression tasks as the dependent variable of regression can be modeled as a multi-valued label using the technique of discretization. However, the extension of classifiers does not result in precise regressors as the discretization error of dependent variable adds up across different trees, creating a less precise forest regressor comparing with the original forest classifier. Here, we utilize Brooks–Iyengar fusion method to create a more-precise random forest regressor.

Definition 8.2 An enhanced random forest regressor is a regressor consisting of a collection of tree-structured regressors $\{r(X, \Theta_k), k \doteq 1, 2, \ldots\}$ where the $\{\Theta_k\}$ are independent identically distributed random vectors, the ith tree gives the predicted value of $r_i(x) \pm z\sigma_i$ at input x (for every $i = 1, 2, \ldots$) with the accuracy of $1 - \mathrm{erf}(z)$, and the predicted value of random forest regressor is equal to $fusion_{BI}(r_1(x) \pm z\sigma_1, r_2(x) \pm z\sigma_2, \ldots)$ for the Brooks–Iyengar fusion function with weights $w_1, w_2, \ldots$.

Considering a sequence of tree regressors $r_1(x), r_2(x), \ldots, r_k(x)$ and assuming that random vector (Y, X) is the superset of randomly constructed training sets for the regressors $r.$, we define the generalization error function $\epsilon(X, Y, r_1, r_2, \ldots, r_n)$ in the following way:

$$\epsilon(X, Y, r_1, r_2, \ldots, r_n) = \sqrt{\overline{\left(r_k(X) - Y\right)^2}_k}$$

where $\overline{\left(r_k(X) - Y\right)^2}_k$ represents the weighted variance of the forest regressor error:

$$\overline{\left(r_k(X) - Y\right)^2}_k = \mathbf{E}\left[\frac{\sum_k w_k \cdot \left(r_k(X) - Y\right)^2}{n}\right]$$

The above value is a convex function of weights $w_1, w_2, \ldots$ and is continuous in the multidimensional weight space. Therefore, it will get a global minimum value for a combination of weights $w_1^*, w_2^*, \ldots$ which we call perfect weights.

According to the strong law of large numbers, we have

$$\lim_{n \to +\infty} \epsilon(X, Y, r_1, r_2, \ldots, r_n) = \sqrt{\lim_{n \to +\infty} \frac{\mathbf{E}\left[\sum_k w_k^* \cdot \left(r_k(X) - Y\right)^2\right]}{n}} < \infty$$

$$(8.2)$$

Relation (8.2) shows that even when the number of trees in a random forest regressor converges to infinity, there will be no overfitting as the error value does not converge to zero.

Since Brooks–Iyengar algorithm is fault tolerant, this makes the random forest regressor more reliable and robust to input data noise and error. In fact, as Eq. (8.2) claims, the overall error value of the random forest depends on the average error values of different trees in the forest. Therefore, even in the case that f tree regressors (say $r_j, j \in F$) are faulty and have high standard errors:

$$\sigma_j \leq \alpha \max_{i \neq j} \sigma_i, \quad \forall j \in F$$

where their standard errors are at-most $\alpha > 1$ times more than the non-faulty tree regressors, the overall error value of the random forest regressor based on Brooks–Iyengar algorithm will be

$$\overline{\sigma.} \leq \frac{\left(n + (\alpha - 1)|F|\right)\sigma}{n} \approx \lim_{n \to +\infty} \frac{\left(n + (\alpha - 1)|F|\right)\sigma}{n} = (\alpha\beta - \alpha + 1)\sigma$$

where $\beta = \frac{|F|}{n}$ is the proportion of faulty tree regressors in the random forest. Therefore, as long as the number of faulty tree regressors is negligible compared with the size of forest regressor ($\beta << 1$), the random forest regressor is fault tolerant.

8.4   Applications in Autonomous Car

Autonomous vehicles are expected to play a key role in the future of urban transportation systems, as they offer potential for additional safety, increased productivity, greater accessibility, better road efficiency, and positive impact on the environment. Research in autonomous systems has seen dramatic advances in recent years, due to the increases in available computing power and reduced cost in sensing and computing technologies, resulting in maturing technological readiness level of fully autonomous vehicles. The objective of this paper is to provide a general overview of the recent developments in the realm of autonomous vehicle software systems. Fundamental components of autonomous vehicle software are reviewed, and recent developments in each area are discussed.

Autonomous vehicles (AVs) are widely anticipated to alleviate road congestion through higher throughput, improve road safety by eliminating human error, and free drivers from the burden of driving, allowing greater productivity and/or time for rest, along with a myriad of other foreseen benefits. The past three decades have seen steadily increasing research efforts in developing self-driving vehicle technology, in part fueled by advances in sensing and computing technologies which have resulted in reduced size and price of necessary hardware. Furthermore, the perceived societal benefits continue to grow in scale along with the rapid global increase of vehicle ownership. As of 2010, the number of vehicles in use in the world was estimated to be 1.015 billion [2], while the world population was estimated to be

6.916 billion [3]. This translates to one vehicle for every seven persons. The societal cost of traffic crashes in the USA was approximately 300 billion USD in 2009 [4]. The financial cost of congestion is continually increasing each year, with the cost estimate for USA reaching as high as 160 billion USD in 2014 [5]. The associated health cost of congestions in USA was estimated to be over 20 billion USD in 2010 from premature deaths resulting from pollution inhalation [6]. While it is uncertain just how much these ongoing costs can be reduced through autonomous vehicle deployment, attempting to curtail the massive scale of these numbers serves as great motivation for the research.

A future with self-driving cars was first envisioned as early as 1918 [7], with the idea even broadcasted over television as early as 1958 [8]. By 1988, Carnegie Mellon's NAVLAB vehicle was being demonstrated to perform lane-following using camera images [9]. Development was accelerated when several research teams later developed more advanced driverless vehicles to traverse desert terrain in the 2004 and 2005 DARPA Grand Challenges [10], and then urban roads in the 2007 DARPA Urban Challenge (DUC) [11]. Research related to self-driving has since continued at a fast pace in academic settings, but furthermore is now receiving considerable attention in industry as well.

As research in the field of autonomous vehicles has matured, a wide variety of impressive demonstrations have been made on full-scale vehicle platforms. Recent studies have also been conducted to model and anticipate the social impact of implementing autonomous mobility-on-demand (MoD) [12]. The case studies have shown that MoD system would make access to mobility more affordable and convenient compared to traditional mobility system characterized by extensive private vehicle ownership.

Autonomous driving on urban roads has seen tremendous progress in recent years, with several commercial entities pushing the bounds alongside academia. Google has perhaps the most experience in the area, having tested its fleet of autonomous vehicles for more than 2 million miles, with expectation to soon launch a pilot MoD service project using 100 self-driving vehicles [13]. Tesla is early to market their work, having already provided an autopilot feature in their 2016 Model S cars [14]. Uber's mobility service has grown to upset the taxi markets in numerous cities worldwide, and has furthermore recently indicated plans to eventually replace all their human driven fleet with self-driving cars [15], with their first self-driving vehicle pilot program already underway [16].

There are several places where automated road shuttles are in commercial operations. Examples include deployments at Rivium Business Park, Masdar City, and Heathrow Airport [13, 17]. The common feature of these operations is that road vehicles are certified as a rail system meaning that vehicles operate in a segregated space [17]. This approach has been necessary due to legal uncertainty around liability in the event of an accident involving an autonomous vehicle. To address this, governments around the world are reviewing and implementing new laws. Part of this process has involved extended public trials of automated shuttles, with CityMobil and CityMobil2 being among the largest of such projects [17].

While the majority of the research contributions discussed in the remaining sections of this article are from academic institutions, it is worth noting that the industrial market interest is also largely responsible for research investigations into certification and validations processes, especially in regards to autonomous car manufacturability and services [18, 19]. These topics are, however, left out of the scope of this survey paper.

Driving in urban environments has been of great interest to researchers due in part to the high density of vehicles and various area-specific traffic rules that must be obeyed. The DARPA Urban Challenge [20], and more recently the V-Charge Project [21] catalyzed the launch of research efforts into autonomous driving on urban road for numerous organizations. Referring to Fig. 8.1, the problem of urban driving is both interesting and difficult because it pushes the research direction to address both increased operating speeds of autonomous vehicles as well increased environmental complexity.

The core competencies of an autonomous vehicle software system can be broadly categorized into three categories, namely perception, planning, and control. Also, vehicle-to-vehicle (V2V) communications can be leveraged to achieve further improvements in areas of perception and/or planning through vehicle cooperation.

Perception refers to the ability of an autonomous system to collect information and extract relevant knowledge from the environment. Environmental perception refers to developing a contextual understanding of environment, such as where obstacles are located, detection of road signs/marking, and categorizing data by their semantic meaning. Localization refers to the ability of the robot to determine its position with respect to the environment.

Planning refers to the process of making purposeful decisions in order to achieve the robot's higher order goals, typically to bring the vehicle from a start location to a goal location while avoiding obstacles and optimizing over designed heuristics.

8.5   Conclusion

The Brooks–Iyengar algorithm is an extension of original Byzantine generals problem, as well as combining the sensor fusion approaches. So Brooks–Iyengar algorithm employs both fault tolerance and fusion characteristics. In the area of newly appeared blockchain applications, Brooks–Iyengar could be used as the consensus algorithm to diversify the application and create new kinds of blockchain systems. In the area of machine learning and deep learning, Brooks–Iyengar algorithm could be used as a basic data fusion algorithm both in training process or simply result fusion. The fault tolerance could enable the machine learning robust to training errors and testing errors. In a nutshell, Brooks–Iyengar algorithm combines both distributed agreement and sensor fusion enables more powerful applications that concern the consensus, fault tolerance, and fusion.

References

1. M.O. Rabin, Randomized byzantine generals, in *24th Annual Symposium on Foundations of Computer Science* (IEEE, 1983), pp. 403–409
2. K. Koscher, A. Czeskis, F. Roesner, S. Patel, T. Kohno, S. Checkoway, D. McCoy, B. Kantor, D. Anderson, H. Shacham, et al., Experimental security analysis of a modern automobile, in *2010 IEEE Symposium on Security and Privacy (SP)* (IEEE, Piscataway, 2010), pp. 447–462
3. A. Eggers, H. Schwedhelm, O. Zander, R.C. Izquierdo, J.A.G. Polanco, J. Paralikas, K. Georgoulias, G. Chryssolouris, D. Seibert, C. Jacob, Virtual testing based type approval procedures for the assessment of pedestrian protection developed within the EU-project IMVITER, in *Proceedings of 23rd Enhanced Safety of Vehicles (ESV) Conference* (Seoul, 2013), pp. 27–30
4. C. Cachin, K. Kursawe, F. Petzold, V. Shoup, Secure and efficient asynchronous broadcast protocols, in *Annual International Cryptology Conference* (Springer, Berlin, 2001), pp. 524–541
5. W.D. Jones, Keeping cars from crashing. IEEE Spectr. **38**(9), 40–45 (2001)
6. L. Breiman, Using adaptive bagging to debias regressions. Technical report, Technical Report 547, Statistics Dept. UCB, 1999
7. N.H. Vaidya, V.K. Garg, Byzantine vector consensus in complete graphs, in *Proceedings of the 2013 ACM Symposium on Principles of Distributed Computing* (Montréal, 2013), pp. 65–73
8. B. Ao, Y. Wang, L. Yu, R.R. Brooks, S. Iyengar, On precision bound of distributed fault-tolerant sensor fusion algorithms. ACM Comput. Surv. **49**, 5 (2016)
9. M.D. Meyer. Crashes vs. congestion–what's the cost to society? AAA Research Report, 2008
10. J.I. Levy, J.J. Buonocore, K. Von Stackelberg, Evaluation of the public health impacts of traffic congestion: a health risk assessment. Environ. Health **9**(1), 65 (2010)
11. A. Miller, J.J. LaViola Jr., Anonymous byzantine consensus from moderately-hard puzzles: a model for bitcoin. http://nakamotoinstitute.org/research/anonymous-byzantine-consensus
12. A. Mostefaoui, H. Moumen, M. Raynal, Signature-free asynchronous binary byzantine consensus with $t < n/3$, O(n2) messages, and O(1) expected time. J. ACM **62**(4), 31 (2015)
13. T. Winter, *The Advantages and Challenges of the Blockchain for Smart Grids*. Delft University of Technology, 2018
14. K. Marzullo, Tolerating failures of continuous-valued sensors. ACM Trans. Comput. Syst. **8**, 284–304 (1990)
15. G.S.Thejas, T.C. Pramod, S.S. Iyengar, N.R. Sunitha, Intelligent access control: a self-adaptable trust-based access control (SATBAC) framework using game theory strategy, in *Proceedings of International Symposium on Sensor Networks, Systems and Security*, ed. by S.V. Nageswara Rao, R.R. Brooks, C.Q. Wu (Springer International Publishing, Cham, 2018), pp. 97–111. https://doi.org/10.1007/978-3-319-75683-7_7
16. M.J. Fischer, N.A. Lynch, M.S. Paterson, Impossibility of distributed consensus with one faulty process. J. ACM **32**(2), 374–382 (1985)
17. S.K. Ramani, S.S. Iyengar, Evolution of sensors leading to smart objects and security issues in IoT, in *International Symposium on Sensor Networks, Systems and Security* (Springer, Berlin, 2017), pp. 125–136
18. B. Ao, Robust fault tolerant rail door state monitoring systems: applying the Brooks–Iyengar sensing algorithm to transportation applications. Int. J. Next Gener. Comput. **8**, 108–114 (2017)
19. M. Buehler, K. Iagnemma, S. Singh. *The DARPA Urban Challenge: Autonomous Vehicles in City Traffic*, vol. 56 (Springer, Berlin, 2009)
20. S. Iyengar, R.R. Brooks, *Multi-Sensor Fusion: Fundamentals And Applications With Software* (Prentice-Hall, Englewood Cliffs, 1997)
21. Y. Amit, D. Geman, Shape quantization and recognition with randomized trees. Neural Comput. **9**(7), 1545–1588 (1997)

Chapter 9
Designing a Deep-Learning Neural Network Chip to Detect Hardware Errors Using Brooks–Iyengar Algorithm

9.1 Motivation

Wireless sensor networks since its inception has penetrated into being used in many modern technologies. Their inexpensive nature and the capabilities they possess have led to them having many applications. Some technologies have a stronger dependence on the sensing capabilities and any ambiguity in the collected data can have catastrophic effects on such systems. To avoid such implications of a faulty sensor, there are many information fusion algorithms that have been proposed. One such algorithm to have found many applications even to this day is the Brooks–Iyengar algorithm which was first proposed in 1996.

In this article, we discuss the implementation details of a deep-learning neural network (DNN) with nodes having the Brooks–Iyengar algorithms running on a chip (or a single chip computer) that can provide efficient and scalable fault tolerant systems. However, implementations of such DNN and their training often have to deal with a trade-off between efficiency and flexibility. It has been seen that the performance of software-based solutions in managing fault tolerance deteriorates with time. On the other hand, hardware-based approaches utilize the parallelism and optimizations and improve the processing capabilities [1].

In this chapter, we adopt the ideas described in [1] in designing a system-on-chip architecture that can implement the Brooks–Iyengar algorithm and thus provide hardware-based fault tolerance. The specific use-case we would like to target is that of autonomous vehicles. The design will try to utilize the inherent parallelism by defining a neural co-processor idealized in [1].

© Springer Nature Switzerland AG 2020
P. Sniatala et al., *Fundamentals of Brooks–Iyengar Distributed Sensing Algorithm*,
https://doi.org/10.1007/978-3-030-33132-0_9

9.2   Design Vision

Our vision in this project is to apply the Brooks–Iyengar algorithm in designing a future deep-learning neural network (DNN) system. This system can presumable be a highly dynamic system with high probability of hardware errors like a self-driving cars or machine learning data centers, etc. Specifically, each neuron in the DNN shall act as a processing element (PE) in the context of the Brooks–Iyengar algorithm. Such a PE will broadcast data to other neurons for processing, and if one such neuron is faulty owing to hardware errors, we expect the system to detect and tolerate the faulty neuron eventually producing correct (or near correct) results as the output of the DNN.

One of the challenges in this design is to keep detection overhead low, because that the algorithm has to be executed in every PE. One of the possible solutions to reduce the overhead is by applying a software-hardware co-design method. The software-hardware co-design will have special circuits in the DNN hardware to execute the Brooks–Iyengar algorithm together with neurons in DNN.

9.3   Introduction

Deep-learning neural networks (DNNs) are a highly popular technology in the modern day because of the power they have to mimic the functionality of the human brain. In specific, these networks have the capability of replicating the important characteristics of the complex brain structure by performing (a) tasks in parallel, (b) computing and processing modularly, (c) ability to learn and thus generalize and relate to the possible outcomes and occurrences. All these features have made these popular and led to their use in many fields. In this chapter we would try to use the multi-layered structure of the DNNs in the design of a network chip that can detect hardware errors and with robust performance using the Brooks–Iyengar algorithm.

DNNs in general can be defined as a set of algorithms that are specialized in identifying patterns. When used in a wireless sensor network (WSN) that constantly collects sensory information, the series of sequential layers replicate the neurons in the brain and along with machine like perception, label and cluster these inputs. They then use these in generating the patterns that are stored as vectors [2].

There has been significant development in the generation of specialized hardware with drastic improvements in the processing capabilities. Many major technology industry giants have been investing time and money in the generation of specialized ASICs [3] which can implement DNNs with tunable parameters so that they perform well in new environments with varying inputs. The design of such systems or networks-on-chip (NoC) has been a topic of research for the implementation of DNNs and ANNs as described in [4–7]. The design flexibility and great computational boost that such chips provide have strengthened their choice in the design of real-time dynamic systems like autonomous vehicles or even datacenters performing complex machine learning operations.

Many existing efforts though assume that the neural networks have been trained reliably through external means and use them as a part of the recall phase [1]. The training process is considered the most complicated and expensive step in the successful working of the system and involves the use of many optimizations. The use of external sources and relying on them for the correct and optimal training is impractical when trying to design a chip suited for generic applications as there can be mismatch of parameters and topology. We thus believe that the software–hardware co-design that can be achieved using an information fusion algorithm like Brooks–Iyengar algorithm will aid in improved training and thus building a highly fault tolerant and precise system that can work well even with real-time and highly time critical inputs.

9.4   System Design and Architecture

Common neuro-chip design and architecture are as shown in Fig. 9.1.

The proposed system for the system/network on chip will constitute of a neural processor that plays a major role in the system. The neural processor for the system can either work independently or be embedded alongside a master microcontroller that handles larger computations and hence working as a co-processor. The neural processor processes information and programs running from a memory block either embedded or externally mounted to the FPGA. The input data for the processor can be training data or instructions or even data that has been broadcasted by other PEs as defined in the Brooks–Iyengar algorithm [8]. The input can be fed either as a control parameter or weight or a vector with training data to the processor's input queue using direct memory access (DMA) enabled devices. The schematic of the design of such an architecture based on the inputs in [1] is as shown in Fig. 9.2. Figure 9.3 shows a repository and the program memory that is interfaced from the external environment or embedded and provided direct access to the DMA or by loading the instructions to a specialized instruction memory from where it can be fetched for processing.

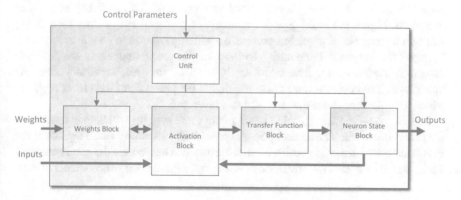

Fig. 9.1 Common neuro-chips

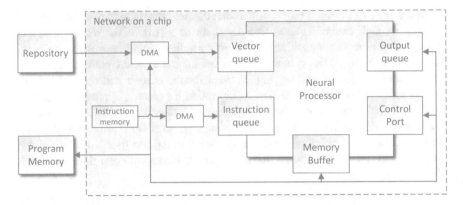

Fig. 9.2 Overall system architecture

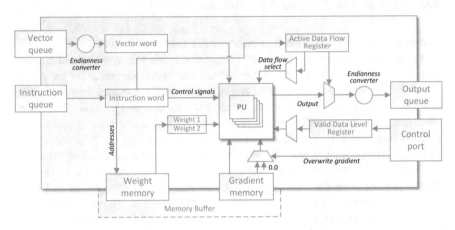

Fig. 9.3 Internal components of a typical neural processor (Adopted from [1])

The neural processor details and constituents are as shown in Fig. 9.3. The brain and main working unit of the processor are made up of entities called processing units (PUs). The PUs are parameterized and provided the capability of parallel execution. Figure 9.4 highlights the internals of a typical PU as described in [1] that will be capable of pipelined processing with a provision to tune it as required. To store the variables temporarily and aid in the processing, there are multiple associated memory units that can be easily accessed for read, write, and other I/O operations. The depicted image also highlights the presence of data level registers which identifies the flows that are used at a given time.

The Brooks–Iyengar algorithm being run in these systems will help in enhancing the precision values. The control port can provide the necessary status signals in the FIFO manner with a configurable bit that the user can use to alter the configurations. These signals are used for execution flow control. Figure 9.4 shows contents of each

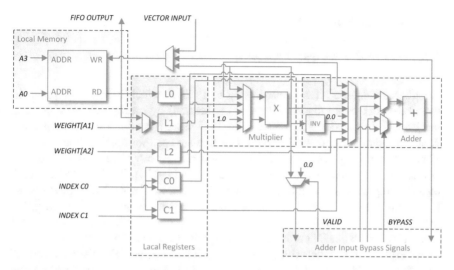

Fig. 9.4 The basic building blocks of a processing unit (can vary based on available resources and application)

PU. Essentially a PU consists of an adder unit, a multiplier unit, a local memory block, and a reduced number of local registers.

9.5   Design for Brooks–Iyengar Algorithm

The Brooks–Iyengar algorithm is executed in every processing element (PE) of a distributed sensor network. Each PE exchanges their measured interval with all other PEs in the network. The "fused" measurement is a weighted average of the midpoints of the regions found. The concrete steps of Brooks–Iyengar algorithm are shown in this section. Each PE performs the algorithm separately.

Algorithm 9.1 specifies the steps through which a specific PE calculates the point estimate and interval estimate assuming that the measurements sent by other PEs are available.

Algorithm 9.1: BI-DNN design approach

Input: The measurement sent by PE_k to PE_i is a closed interval
Output: The output of PE i includes a point estimate and an interval estimate

1: Add the measurements sent to S_i from all other S_j where $1 <= j <= N$
2: Multiply the sum by weights W_i
3: Divide by 2 and obtain the numerator of point estimate
4: Add of all weights in that interval
5: Divide the numerator obtained in step 3 by the sum in step 4
6: Obtain the point estimate for S_i. This is the required functional computation
7: Calculate the interval estimate

9.6 Similar Attempts

9.6.1 Google: Tensor Processing Unit (TPU)

Google has released a tensor processing unit (TPU) which is built to support the neural network-based machine learning applications by specialized ASIC chips. It has been in the market since 2016. The TPU plays the role of an AI accelerator in combination with other CPU and GPUs that are used for computation activities. There are various generations of TPUs with the most recent being the Edge TPU released in July 2018. According to Google Cloud [9, 10], the TPU ASIC is built on a 28 nm processor and runs at 700 MHz. It consumes about 40 W of power when it is in the running state. It is modeled as a card that fits into a traditional SATA hard-disk slot for easy drop-in installation. TPU is connected with its host via a PCI e-gen 136 3 × 16 bus that provides 12.5 Gbps of effective bandwidth.

Excerpts from an article titled "First in-depth look at Google's TPU architecture" from the Next Platform states "We did a very fast chip design. It was really quite remarkable. We started shipping the first silicon with no bug fixes or mask changes. Considering we were hiring the team as we were building the chip, then hiring RTL (circuitry design) people and rushing to hire design verification people, it was hectic" [10].

According to them, each neuron has the capability to:

- Combine the input data (x) with the corresponding weights (w) using a multiplier unit to be able to precisely depict the strength of the signal.
- Collect and thus aggregate the values of the neuron states and merge it into a single value.
- Use activation functions like tanh, Sigmoid, ReLu that will be able to modulate the synthesized neuron activity.

9.6.1.1 Example

- **Inputs:** x_1, x_2, x_3
- **Weights:** $w_{11}, w_{12}, w_{13} \ldots$
- **Outputs:** $Y_1 = f(w_{11}x_1 + w_{12}x_2 + w_{13}x_3)$ and $Y_2 = f(w_{21}x_1 + w_{22}x_2 + w_{23}x_3)$

In the example, if we have 3 inputs and 2 neurons with a fully connected single layer neural network, you have to execute six multiplications between the weights and the inputs and add up the multiplications in 2 groups of 3. This sequence of multiplications and additions can be written as a matrix multiplication. The outputs of the matrix multiplication are then processed by an activation function.

9.6.2 IBM AI Research

The research team at IBM has been working on developing hardware architectures and devices that can be used in realizing great processing speeds. Their research also focuses on a transition from narrow AI to broad AI [11] that can support machine intelligence, in-memory computing, and other features on complex workloads. Some of their designs have been breakthrough ideas in taking AI further and even to the edge through ultra-precision training of DNNs [12–14].

9.6.3 Intel's Nervana Neural Network Processors

One of the major chip manufacturing company, Intel has launched their series of processors that can perform the functionality of neural networks. Such processors called the Nervana neural network processors (Intel NNPs) are touted to enable the seamless use of distributed learning algorithms and systems. According to Intel, these processors can help transform the deep learning-based reasoning into data for global knowledge by harnessing the benefits that AI has to offer [15].

9.7 Conclusion

The Brooks–Iyengar DNN architecture thus designed, can detect and tolerate the faulty neuron and aid in producing near correct results as the output of the DNN. One of the challenges though is to keep the detection overhead as low as possible because the algorithm has to be executed per PE. This overhead can be reduced by applying a software–hardware co-design technique, i.e., by having a specialized circuit designed with neurons in the DNN.

References

1. R.J. Aliaga, R. Gadea, R.J. Colom, J.M. Monzó, C.W. Lerche, J.D. Martínez, System-on-chip implementation of neural network training on FPGA. Int. J. Adv. Syst. Meas. 2(1), 2009 (2009)
2. A beginner's guide to neural networks and deep learning. Skymind. https://skymind.ai/wiki/neural-network. Accessed 10 Sept 2019
3. K. Ovtcharov, O. Ruwase, J.-Y. Kim, J. Fowers, K. Strauss, E.S. Chung, Accelerating deep convolutional neural networks using specialized hardware. Microsoft Res. Whitepaper 2(11), 1–4 (2015)
4. D.K. McNeill, C.R. Schneider, H.C. Card, Analog CMOS neural networks based on Gilbert multipliers with in-circuit learning, in *Proceedings of the 36th Midwest Symposium on Circuits and Systems*, vol. 2 (1993), pp. 1271–1274

5. R.J. Aliaga, R. Gadea, R.J. Colom, J.M. Monzó, C.W. Lerche, J.D. Martinez, A. Sebastiá, F. Mateo, Multiprocessor SoC implementation of neural network training on FPGA, in *2008 International Conference on Advances in Electronics and Micro-electronics (ENICS)* (2008), pp. 149–154
6. A.W. Savich, M. Moussa, S. Areibi, The impact of arithmetic representation on implementing MLP-BP on FPGAs: a study. IEEE Trans. Neural Netw. **18**(1), 240–252 (2007)
7. T. Theocharides, G. Link, N. Vijaykrishnan, M.J. Irwin, V. Srikantam, A generic reconfigurable neural network architecture as a network on chip, in *Proceedings of the IEEE International SoC Conference* (2004), pp. 191–194
8. R.R. Brooks, S.S. Iyengar, Robust distributed computing and sensing algorithm. Computer **29**(6), 53–60 (1996). https://doi.org/10.1109/2.507632. ISSN 0018-9162
9. Cloud Tensor Processing Units (TPUs), Cloud TPU. Google Cloud. Google. https://cloud.google.com/tpu/docs/tpus. Accessed 11 Sept 2019
10. An in-Depth Look at Google's First Tensor Processing Unit (TPU). Google Cloud Blog. Google. https://cloud.google.com/blog/products/gcp/an-in-depth-look-at-googles-first-tensor-processing-unit-tpu. Accessed 11 Sept 2019
11. AI hardware, AI Hardware—IBM Research AI. https://www.research.ibm.com/artificial-intelligence/hardware/#papers. Accessed 5 June 2018
12. W. Kim, R.L. Bruce, T. Masuda, G.W. Fraczak, N. Gong, P. Adusumilli, S. Ambrogio et al., Confined PCM-based analog synaptic devices offering low resistance-drift and 1000 programmable states for deep learning, in *2019 Symposium on VLSI Technology* (IEEE, Piscataway, 2019), pp. T66–T67
13. H.-Y. Chang, P. Narayanan, S.C. Lewis, N.C.P. Farinha, K. Hosokawa, C. Mackin, H. Tsai, S. Ambrogio, A. Chen, G.W. Burr. AI hardware acceleration with analog memory: micro-architectures for low energy at high speed. IBM J. Res. Dev. **63**(6), 8–11 (2019)
14. C. Ríos, N. Youngblood, Z. Cheng, M. Le Gallo, W.H.P. Pernice, C.D. Wright, A. Sebastian, H. Bhaskaran. In-memory computing on a photonic platform. Sci. Adv. **5**(2), eaau5759 (2019)
15. Nervana Neural Network Processor. Intel AI. https://www.intel.ai/nervana-nnp/. Accessed 1 Dec 2019

Chapter 10
Ubiquitous Brooks–Iyengar's Robust Distributed Real-Time Sensing Algorithm: Past, Present, and Future

This paper primarily benevolences a two-decade longstanding and most influential Brooks Iyengar's hybrid algorithm known as robust distributed computing and sensing algorithm published in IEEE computing in 1996. The algorithmic architecture establishes foundation principles across various real-time operating systems, application areas, and fault tolerant schemes. The key highlight of the algorithm is in the context of sensor applications growing interest in real-time processing and enhancing their fault tolerance characteristics of the whole system by exploiting the redundancy. The crucial contribution of the algorithm is majorly found in enhancing the features of MINIX real-time operating system, the hybrid architecture, and scalability of the algorithm is proficient enough to encounter the unreliable and distributed sensors data using the Byzantine [1] agreement and distributed decision-making process methods. This article emphasizes on inclusion and adoption of most persuasive long running Brooks Iyengar's algorithm in MINIX real-time operating system and their recent enhancement of incorporating the fault tolerant schemes. Further, the richness of algorithm has acclaimed by millions of vivid category of users around the globe in their research and computational tasks. USC/ISI. Additionally, the scalability of algorithm proved to be beneficial to other domains like cyber physical systems [3], robot merging, high-performance computing, reliability of software and hardware appliance, and artificial intelligence systems. In this paper, we attempt to showcase the use cases and real-time deployments of Brooks–Iyengar's algorithm in various aspects of distributed computational and sensor world.

The following chapter is reprinted by permission from Dr. Latesh Kumar K.J, MTS-Programmer, Computer Science and Engineering, Cloud and Cyber Security Consultant, latesh@sit.ac.in.

© Springer Nature Switzerland AG 2020
P. Sniatala et al., *Fundamentals of Brooks–Iyengar Distributed Sensing Algorithm*,
https://doi.org/10.1007/978-3-030-33132-0_10

175

10.1   Introduction: Brooks–Iyengar Algorithm

The current internet world consist numerous automated systems that must communicate with dynamic atmospheres. Because these environments are undeterminable, the systems depend on sensors for the delivery of information in order to perform the computation. The sensors are unenviable interface across computer systems and internet world (real world) for the communication of data. The smart intelligence development and deployment into these automated control systems is arduous because of limited sensor accuracy, and the noise in data readings recurrently corrupts the accuracy of data. The Brooks–Iyengar algorithm is widely known as Brooks–Iyengar hybrid algorithm [4], this algorithm acclaimed for its betterment in the accuracy of interval measurement engaged by a distributed sensor network. The key merit of this algorithm is even with the faulty sensor [5] presence the smart intelligence of sensor network swaps the measured value and precision value at each node with every peer node. Further provides accuracy in the measured range value for all the nodes of network. The resilient point of algorithm is that it is a fault tolerant and distributed, it does not malfunction even if some sensors transmit faulty data, because of this key feature it is used as sensor fusion method. Further, accuracy and precision assurance are proved in 2016 [6]. The algorithm Brooks–Iyengar integrates with Byzantine agreement with sensor fusion to control the presence of noise in sensor data. The algorithm is designed to channel the space between Byzantine fault tolerance [1] and sensor fusion. Further this algorithm is identified as the first algorithm to amalgamate these dissimilar fields. Principally, it syndicates algorithm of Dolev's [7] for an imprecise contract with fast convergence algorithm (FCA) by Mahaney and Schneider's. The core of algorithm pretends processing elements (PEs) as N and t of them are assumed to be faulty and perform malevolently. It accepts both real and unreal values with noise and uncertainty. However, the output produced by the algorithm is real value with appropriate stipulated accuracy. The algorithm is further customized to resemble Crusader's Convergence Algorithm (CCA) [8], this adoption increases the bandwidth requirement in processing of algorithm. The benefits of algorithm are wide spread across domain like high-performance computing [9], distributed control, software reliability, and real-time MINIX operating systems. The use of algorithm is not restricted to specific domains and applications we have cited. In general, all floating-point computations produce inaccuracy and this varies from machine to machine computing. This hybrid algorithm offers increased scientific consistency on a distributed system encompassing assorted components. This offers a novel method of resistance in rounding and skewing the errors generated by hardware limitations. Today's cloud based software development and customer requirements are inconsistent, the cloud based various services like software, platform, data, network, security, and recovery are different in domain but targets to produce a common service to customer. In these environments, faults tolerance and accuracy of services must be assured from end to end terms. The Brooks–Iyengar's algorithm is useful and effective in these instances by achieving robust and distributed accuracy because of the novel intelligence of algorithm. The cluster computing involves the critical data and service modules that are important systems which demand the additional strength and accurateness. Regrettably, data, service, and security are compromised often between them, nevertheless the usage of algorithm increases without sacrificing the accuracy of data, service, and security. The fault tolerance mechanism defined in the algorithm is highly beneficial in both active and passive cluster computing in primary and disaster computing sites. With this algorithm, robust distributed computing applications can be developed and deployed seamlessly. Today's world is full of Internet of Things (IoT) and cloud based services in which sensors are vital part computing systems. The amount of data communication across current dynamic environments is leading to errors, mechanical failures, and uncertainties in sensors. To avoid this, the backup mechanism is plugged but fault tolerance and accuracy cannot be managed. Hence, the Brooks–Iyengar's algorithm has lower bound and upper bounds, and using this technique inaccuracy is dealt smartly and specifically. The algorithm is not limited to a specific domain neither restricted to set of computing and hence the wide spread smartness of algorithm is enchasing from last 20 years by various researchers, computing labs, and university training projects. In illustrating an example, a robust fault tolerant rail door state monitoring system is developed using the Brooks–Iyengar sensing algorithm to transportation applications [10], in this paper the author Buke Ao clearly listed the implementations of Brooks–Iyengar algorithm in variety of redundancy applications by various research studies [11, 12]. The key identification is Brooks–Iyengar algorithm has prominently extended seamlessly by connecting the legacy and edge cutting trends of technology variants in software applications and hardware control systems in cloud and non-cloud systems [10]. The major contribution of algorithm is identified with relevance to Linux and Android operating systems effectively. Truly, tons of various software applications and hardware control systems have encapsulated the Brooks–Iyengar algorithm to offer fault tolerant fusion data across billions of users accessing the various services through internet and other sources of digital media. Further, algorithm indirectly benefits across to 99% of world's top supercomputers and 89% of smart phones and millions of end users around the globe in various computing ways.

10.2   Real-Time MINIX Operating System

The Real-Time MINIX operating system is an enhanced version of MINIX operating system; this was originally programmed by scientist Andrew Tanenbaum for teaching operating system on $\times 86$ computer system. The research study and implementation by the author Gabriel Wainer [13, 14] changed the MINIX operating system to support RT-processing named it "RT-MINIX" by adopting Brooks–Iyengar algorithm in the areas like scheduling algorithms selection, scheduled queues, real-time metrics collection, and fault tolerant systems. Before we explain the deep impact of Brooks–Iyengar algorithm on MINIX operating system, we intend to detail about the MINIX operating system. The detailed understanding on

MINIX operating system sets a platform for understanding the Real-Time MINIX (RT-MINIX) for various applications services and control systems. The MINIX operating systems drivers, a user and system specific server runs on highest level on the miniature kernel architecture. The SYS and CLOCK are the two major tasks responsible to support the user-mode sections of the operating system at higher levels. Apart from this programming the MMU, CPU, interrupt handling, and IPC are the other privileged operations of the MINIX kernel. Just like any other operating system the functionalities like file system (FS), memory management (MM), user management (UM), and process management (PM) are offered by MINIX. The key and unique feature of MINIX over other operating system is stealthily this RS server monitors all the device drivers and various servers inside the operating systems at all time and fix it automatically when any failure is noticed.

All system calls are focused blatantly by system libraries to right server to manage the kernel communication. Let us consider a user requests a process to run an application task, usually a process is initiated by fork () library function when the process manager approves it by verifying with the memory manager on the process slots availability. If any slot available, then process manager instructs the kernel to produce a copy of the process, all this happens transparently without the notice of the user application task. Just like UNIX the MINIX kernel is responsible of managing hardware and device drivers. This involves process scheduling, interrupt management, memory, device I/O, and CPU management. The two major core components of kernel space SYS and CLOCK are explained here because in later sections we illustrate how the Brooks–Iyengar algorithm is seminal for the RT-MINIX enhancement. The SYS control is known as system task, this is vital for all kernel mode operations for the device drivers and key heartbeat channel for user segment servers. Any user request to process internally sends a signal to kernel through the library function; each request is passed to SYS. There are various categories based on the SYS management on kernel calls, to copy data between process SYS calls SYS-VIRCOPY and to configure an alarm SYS-SETALARAM etc. Few new systems call defined in the MINIX are listed below in Fig. 10.1.

The second core object is CLOCK by which kernel manages the process scheduling, timers, cron services, hardware clock, and CPU usage. The interrupt handler will initiate a timer moment when a MINIX system is power on, since then each tick is countered using this interrupt timer. In general, the cooperating servers are created by modulating the operating system; the native MINIX operating system allows third party device driver un-trusted code to run and communicate with kernel,

Kernel Call	Purpose
SYS_VIDEVIO	Read or Write a vector of I/O ports
SYS_VIRCOPY	Safe copy between address spaces
SYS_GETINFO	Get a copy of kernel information

Fig. 10.1 Privileged SYS calls to kernel

the MINIX is smart and manages the spreading of failures. A tight coupling of devices and library functions are created to intact the seamless communication in the low-level kernel operations. In this paper we are describing MINIX operating system in deep to prove how the Brooks–Iyengar algorithm is influencing the fault tolerant and robust distributed control systems in RT-MINIX.

10.3 Influence of Brooks–Iyengar Algorithm

Brooks and Iyengar's; the name is all over the globe from last two decades, this algorithm is considered to be the all-time best robust algorithm for precision, fault tolerance, and isolation of errors across software applications and hardware control systems. In this segment we narrate the Brooks–Iyengar algorithm's influence in various domains like MINIX operating system, sensor networks, software application development, real-time extensions, virtualization, physical cyber systems, and cloud computing. To begin with we explain the development and deployment of a distributed sensing algorithm that has major influence on computing systems.

10.3.1 Brooks–Iyengar's Algorithm on MINIX Operating System

The MINIX operating system powered by Tanenbaum's [15] was enhanced to Real-Time (RT) MINIX operating system and services by Wainer and it is identified as RT-MINIX [13, 14]. Further, more recent features were added to shape up an academic real-time operating system called as MINIX v2. This architecture of design was proposed to train the RTOS with few major topics:

- System architecture
- Handling interrupt
- Process management
- Scheduling of process
- Fault tolerance
- Isolation of errors

The research study by Gabriel Weiner also mentioned that many other control systems, computer application, and real-time systems are created based on the services offered by Brooks–Iyengar algorithm. The services provided by the algorithm on real-time systems, computer applications, and various systems are vaguely different from traditional systems and they are unique and different from the native operating system.

Figure 10.2 describes novel features added to MINIX operating system by Gabriel and Team in creation of RT-MINIX by using the intelligence of Brooks–Iyengar algorithm. The programming of the MINIX source code [13] was dedicated

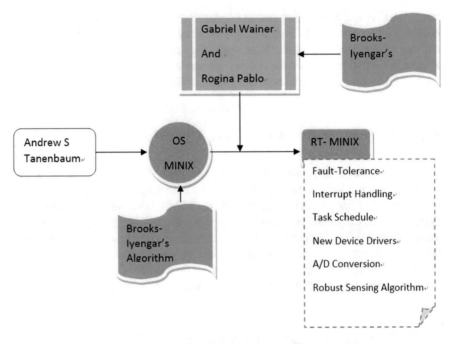

Fig. 10.2 Block structure view of Brooks–Iyengar algorithm influence

to provide the real-time controls on various services. Many real-time services were added, to begin with rate-monotonic scheduling [16], earliest deadline first processor, and fault tolerance are programmed. To make these new changes in the source code of the kernel, the code flow and data structures are slightly modified based on the new updates, specifically, sensor, timers, schedule, and criticality. Further, to adapt live-tasks with interactive CPU bound tasks a multi-queue is developed. The below listed data structure is modified in lieu of RT-MINIX evaluation (Fig. 10.3).

All these changes are tested with various feasibility of MINIX for the real-world challenges for real-time development. Numerous works were done using Brooks–Iyengar robust distributed computing algorithm from the testing of novel scheduling procedures to kernel alterations. In the meantime new version of MINIX was released and hence to sync the RT-MINIX version with MINIX version, some changes were made. The analog to digital conversion [17], in this update the target was to acquire data from analogic environment as many real-time systems are employed to handle the real process like chemical and a production line. In this requirement the Brooks–Iyengar algorithm's sensor management intelligence is effectively used for sensing the real-world data, to control the noise and to manage the faulty sensors. The interface used for game ports was used to provide the signals from the sensors, this was considered and a device driver for port is developed. The changing environment relies on poor performance of integral systems of RT-MINIX

Data Structure	Parameters	Description
struct rt_globstats { int actpertsk; int actapetsk; int misperdln; int misapedln; int totperdln; int totapedln; int gratio; clock_t idletime; };	Acrta_petsk, Act_pertsk Mis_perdln, Mis_apedln Tot_pertsk, Tot_petsk Gratio Idletime	Period and aperiodic real time tasks, total running tasks. Total missed deadlines Total real-time task scheduled instance Guarantee ratio between deadline and instances Computing Time in Second (clock Tick)

Fig. 10.3 Updated MINIX data structure using the Brooks–Iyengar algorithm

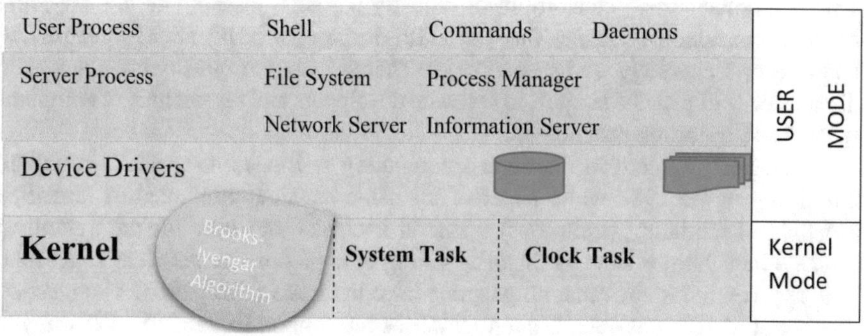

Fig. 10.4 MINIX kernel and Brooks–Iyengar algorithm

with novel techniques. The Brooks–Iyengar's algorithm adopted fast convergence algorithm (FCA) [18] to increase the convergence ratio. According the Pablo Ragina and Gabrile Weiner, the algorithm [19] is used extensively to extend RT-MINIX with possibility of several sensors from a fault tolerance perception. At the outset, the complete coding was performed based on all the four algorithms of hybrid Brooks–Iyengar. The next immediate phase was to integrate the smart capability to make use of the real-time data, to do this four potentiometers were used to sense the signals/data from analogic inputs from the joystick port. These sensor positions are arranged with actual positions for a simulation based robotic arm. An accurate and precise functionality of algorithm was noticed by providing an exclusive value from the simulated sensors in spite of faulty, at the same time users were offered open chance to modify the data by varying the potentiometers. At last, all the updated code is test for various feasibilities and real-time constraints and then the novel algorithm intelligence is united into MINIX kernel. Figure 10.4 shows the RT-MINIX Kernel and new feature additions with Brooks–Iyengar algorithm.

The software developers were given a set of functions to work with intellectual sensors, using these it was possible to generate many new services and devices like/dev/js0 and after that smart sensors were able to read data in the presence of faulty sensors. Once the operating system is enhanced with RT services, the demand ascended for various computing tools and applications. The Brooks–Iyengar's algorithm needed a test on the novel techniques applied on kernel, in order to evaluate the data structure through vivid system and library calls.

10.3.2   Case Study: Open MPI + Virtualization

The Brooks–Iyengar algorithm was further implemented on Linux using the Open-MPI [20], this is an open source project created to pass message through interface. This is a collaborative consortium of industry partners, research community, and groups of academic. Hence, the OpenMPI is powerful and smart because the knowledge, technology, and resources are shared from various communities. The libraries of MPI provide support to software developers and researchers of computer science and operating researchers.

A classical problem in distributed computing is Byzantine generals problem, introduced in the 1982 white paper of the same name. It attempted to formalize a definition for faulty (traitorous) nodes in a cluster and how for the system to mitigate it. Solutions such as majority voting or signed messages were suggested. Majority voting requires that all generals have the same information, a suggestion that is not always possible. Signed messages are a good to verify that it was the correct node in communication, even if it does not verify that the content itself is correct. Both are good suggestions, but it would be more interesting to have an algorithm that can survive a traitorous order every now and then. Enter the Brooks–Iyengar algorithm as an attempt to solve this problem. This algorithm uses sensor fusion to mathematically eliminate the faulty sensors. In short, this is achieved by taking measurements over an interval, the measured interval is then shared between all sensors on the network. The fusion step then happens, by creating a weighted average of the midpoints of all the intervals. At this point you can eliminate any sensors with high variance or use heuristics to choose the reliable nodes. It runs in $O(N \log N)$ time and can handle up to $N/3$ faulty sensors (Fig. 10.5).

To conclude the obtained results it is better to consider the dumb average because, noise generated from real and faulty sensors are from undeviating distribution. If the algorithm has not performed better then the noise would have been twisted and tremendous in one direction causing the red line curve aggressive over the green line. Overall, the algorithm is very difficult to implement as there were no framework/library and demands precision of coding and adequate infrastructure to achieve best results. The results proved that Brooks–Iyengar's algorithm is smart and scalable across various domains like cyber physical system.

OpenMPI Methods	Description
Isend and Ireceive	Non-blocking sends and receives were used to communicate from sensor to sensor. In order to process the data each sensor needs the data from every other sensor in the network. This means there are a worst case of N^2 messages being passed at any given point. Due to this large number, it is best to use non-blocking communications.
Barrier	This acts as a sync step for all sensors. Barrier merely acts as a join for processes in the context of OpenMPI. It is very useful for a simulation as this to stop one process from being a front runner.
Broadcast	Seeing as this is a timed excution program, it is necessary for each sensor to kill itself after a fixed period of time. However, it is possible for one process to keep running if it gets to the check before all the others. To get around this one thread was designated with the resposibility to check the runtime, and then broadcasted the result to all others.

Fig. 10.5 OpenMPI methods implemented using Brooks–Iyengar

10.4 Conclusion

In this article the acceleration, effectiveness, and liveliness of two decade old Brooks–Iyengar algorithm is illustrated. Since today's technology does not guarantee success and safety in all situations, the Brooks–Iyengar algorithm can significantly improve the fault tolerance of systems by providing a greater margin of safety for operations. This algorithm provides the robust implementation and seamless scalability under faulty sensor conditions for various domains. Finally, the algorithm "Stand the Test of Times" from last two decades and hope it continues the successful journey further.

Acknowledgements Authors of this book would like to thank Dr. Kumar as a prominent researcher on Storage—Cloud—Cyber-security—Protocol Engineering for the contribution to the chapter.

References

1. D. Dolev, The Byzantine generals strike again. J. Algorithms **3**(1), 14–30 (1982)
2. Penn State University, Reactive Sensor Networks, AFRL-IF-RS-TR-2003-245, Directorate, Public Affairs Office (IFOIPA) and is releasable to the National Technical Information Service (NTIS), Defense Advanced Research Laboratory, 2013
3. J. Park, R. Ivanov, J. Weimer, M. Pajic, S.H. Son, I. Lee, Security of cyber-physical systems in the presence of transient sensor faults. J. ACM Trans. Cyber-Phys. Syst. **1**(3), 15 (2017). https://doi.org/10.1145/3064809
4. R.R. Brooks, S.S. Iyengar, Robust distributed computing and sensing algorithm. Computer **29**(6), 53—60 (1996). https://doi.org/10.1109/2.507632. ISSN 0018-9162. Archived from the original on 2010-04-08. Retrieved 2010-03-22

5. M. Ilyas, I. Mahgoub, *Handbook of Sensor Networks: Compact Wireless and Wired Sensing Systems* (CRC Press, Boca Raton, 2004). pp. 254, 33-2 of 864. bit.csc.lsu.edu. ISBN 978-0-8493-1968-6. Archived from the original (PDF) on June 27, 2010. Retrieved March 22, 2010
6. B. Ao, Y. Wang, L. Yu, R.R. Brooks, S.S. Iyengar, On Precision bound of distributed fault-tolerant sensor fusion algorithms. ACM Comput. Surv. **49**(1), 5 (2016). https://doi.org/10.1145/2898984. ISSN 0360-0300
7. L. Lamport, R. Shostak, M. Pease, The Byzantine generals problem. ACM Trans. Program. Lang. Syst. **4**(3), 382–401 (1982). CiteSeerX 10.1.1.64.2312. https://doi.org/10.1145/357172.357176
8. D. Dolev, et al., Reaching approximate agreement in the presence of faults. J. ACM **33**(3), 499–516. CiteSeerX 10.1.1.13.3049. https://doi.org/10.1145/5925.5931. ISSN 0004-5411. Accessed 23 March 2010
9. S. Mahaney, F. Schneider, Inexact agreement: accuracy, precision, and graceful degradation, in Proceedings of Fourth ACM Symposium Principles of Distributed Computing (1985), pp. 237–249. CiteSeerX 10.1.1.20.6337. https://doi.org/10.1145/323596.323618. ISBN 978-0897911689
10. B. Ao, Robust fault tolerant rail door state monitoring systems: applying the Brooks–Iyengar sensing algorithm to transportation applications. Int. J. Next Gener. Comput. **8**(2), 108–114 (2015)
11. V. Kumar, Computational and compressed sensing optimizations for information processing in sensor network. Int. J. Next Gener. Comput. **3**(3), 1–5 (2012)
12. B. Ao, Y. Wang, L. Yu, R.R. Brooks, S.S. Iyengar, On precision bound of distributed fault-tolerant sensor fusion algorithms. ACM Comput. Surv. **49**(1), 5:1–5:23 (2016)
13. P.J. Rogina, G. Wainer, New real-time extensions to the MINIX operating systemİ, in *Proceedings of 5th International Conference on Information System Analysis and Synthesis (IASS '99)* (1999)
14. G.A. Wainer, Implementing Real-Time services in MINIX. ACM Oper. Syst. Rev. **29**(3), 75–84 (1995)
15. S. Tanenbaum Andrew, S. Woodhull Albert, *Sistemas Operativos: Diseno e Implementacion*, 2nd edn. (Prentice Hall, Englewood Cliffs, 1999). ISBN 9701701658
16. K. Chakrabarty, S.S. Iyengar, H. Qi, E.C. Cho, Grid coverage of surveillance and target location in distributed sensor networks. IEEE Trans. Comput. **51**(12), 1448–1453 (2002)
17. B. Krishnamachari, S.S. Iyengar, Distributed Bayesian algorithms for fault-tolerant event region detection in wireless sensor networks. IEEE Trans Comput. **53**(3), 241–250 (2004)
18. S. Mahaney, F. Schneider, Inexact agreement: accuracy, precision, and graceful degradation, in *Proceedings of Fourth ACM Symposium Principles of Distributed Computing* (ACM Press, New York, 1985), pp. 237–249
19. R. Brooks, S. Iyengar, Robust distributed computing and sensing algorithm. IEEE Comput. **29**(6), 53–60 (1996)
20. Warrenedgar, An implementation of the Brooks–Iyengar algorithm using OpenMPI (2019). https://github.com/warrenedgar/brooks-iyengar

Correction to: Fundamentals of Brooks-Iyengar Distributed Sensing Algorithm

Correction to:
P. Sniatala et al., *Fundamentals of Brooks-Iyengar Distributed Sensing Algorithm,* **https://doi.org/10.1007/978-3-030-33132-0**

This book was inadvertently published with an incorrect affiliation of the author "M. Hadi Amini". The affiliation has been corrected now and it reads as follows:

M. Hadi Amini
School of Computing and Information Sciences, Florida International University, Miami, USA.

The updated online version of the book can be found at
https://doi.org/10.1007/978-3-030-33132-0

Appendix

In this appendix, we list selected Dr. Iyengar's sensor-related books (9), peer-reviewed journal papers (58), and refereed conference papers/book chapters (104):

Authored/Edited Books

1. S. S. Iyengar, Kianoosh G. Boroojeni, N. Balakrishnan, "Mathematical Theories of Distributed Sensor Networks", Springer Verlag, pp 240, December 2014.
2. S.S.Iyengar. N. Parameshwaran et. al., "Fundamentals of Sensor Network Programming: Application and Technology ", John Wiley and sons and IEEE Press, November 2010, pp. 352
3. Chakrabarty and S.S. Iyengar, "Scalable Infrastructure for Information Processing in Distributed Sensor Networks", Springer-Verlag London Ltd, June 2005, pp. 252.
4. R.R. Brooks and S.S. Iyengar "Multi Sensor Fusion: Fundamentals and Applications with Software", Prentice Hall Publication Co., New Jersey 07458 (October 1997), pp. 488.
5. S.S. Iyengar, L. Prasad and Hla Min, "Advances in Distributed Sensor Integration: Applications and Theory", Prentice-Hall, New Jersey, (1995), pp. 273.
6. S.S. Iyengar, R.R. Brooks (EDS), "Distributed Sensor Networks: Image and Sensor Signal Processing", 2nd Edition, CRC Press, Chapman & Hall Books, September, 2012.
7. S.S. Iyengar, R.R. Brooks (EDS), "Distributed Sensor Networks: Sensor Networking and Applications", 2nd Edition, CRC Press, Chapman & Hall Books, September, 2012.
8. V.K. Prasanna, S. S. Iyengar, Paul Spirakis, Matt Welsh (EDS), "Distributed Computing in Sensor Systems", Proceedings of IEEE International Conference on DCOSS, pp. 420, 2005.

© Springer Nature Switzerland AG 2020
P. Sniatala et al., *Fundamentals of Brooks–Iyengar Distributed Sensing Algorithm*,
https://doi.org/10.1007/978-3-030-33132-0

9. S.S. Iyengar and R.R. Brooks (EDS), "Distributed Sensor Networks", Taylor and Francis/CRC Press, Inc. December 2004, pp. 1120.

Peer-reviewed Journals

1. Buke Ao, Yongcai Wang, Richard Brooks, Iyengar S.S., and Yu Lu "On Precision Bound of Distributed Fault-Tolerant Sensor Fusion Algorithms," Vol. 49, Issue 1, ACM Computing Surveys, 2016.
2. Hien Nguyen, Ebtissam Wahman, Niki Pissinou, S. S. Iyengar, Kia Makki, "Mobile Learning Object Authoring Tool and Management System for Mobile Ad Hoc Wireless Networks," International Journal of Communication Systems, Vol. 28, Issue 17, pp. 2180–2196, 2015.
3. Yongcai Wang, Lei Song, S.S. Iyengar, "An Efficient Technique for Locating Multiple Narrow-band Ultrasound Targets in Chorus Mode", IEEE Journal on Selected Areas in Communications, Vol. 33, No. 11, pp. 2343–2356, Nov. 2015.
4. Neeta Trivedi, S. Sitharama Iyengar, N. Blakrishnan, "Energy-Efficient, Delay-Constrained, QoS-aware Broadcast for Cooperative Wireless Sensor Networks," Int. Journal of Sensor Networks, Vol. 16, No. 2, pp. 114–126, Nov. 2014.
5. Jaime Ballesteros, Bogdan Carbunar, Mahmudur Rahman, Naphtali Rishe, and S.S. Iyengar, "Towards Safe Cities: A Mobile and Social Networking Approach," IEEE Transaction on Parallel and Distributed Systems, Vol. 25, No. 9, pp. 2491–2462, Sep. 2014.
6. Scott C.-H. Huang, Hsiao-Chun Wu, and Sundaraja Sitharama Iyengar, "Multisource Broadcast in Wireless Networks," IEEE Transactions on Parallel and Distributed Systems, Vol. 23, Issue 40, pp. 1908–1914, Oct. 2012.
7. Youngki Lee, S.S. Iyengar, Chulhong Min, Younghyun Ju, Seungwoo Kang, Taiwoo Park, Jinwon Lee, Yunseok Rhee, and Junehwa Song, "MobiCon-A Mobile Context-Monitoring Platform", Communications of the ACM, Vol. 55, No.5, pp. 54–65, March 2012.
8. Lu Lu, Hsiao-Chun Wu, Kun Yan, S.S.Iyengar, "Robust Expectation- Maximization Algorithm for Multiple Wide-Band Acoustic Source Localization in the Presence of Non-Uniform Noise Variances" IEEE Sensors Journal, Vol. 11, No.3. March 2011.
9. Zakiya S. Wilson, Sitharama S. Iyengar, Su-Seng Pang, Isiah M. Warner, and Candace Luces, "Increasing Access for Economically Disadvantaged Students: The NSF-CSEM & S-Stem Programs At Louisiana State University", Journal of Engineering Education (JEE), Vol. 21, Issue 5, pp. 581–587, November 2011.
10. Noureddine Boudriga, Mohamed Hamdi, S.S. Iyengar, "Coverage Assessment and Target Tracking in 3D Domains", Journal of Sensors, Vol. 11, Issue 10, pp. 9904–9927, April 2011.

11. Jren-Chit Chin, Nageshwara S. V. Rao, David K. Y. Yau, Mallikarjun Shankar, Yong Yang, Jennifer C. Hou, Srinivasagopalan Srivatsan and S.S.Iyengar, "Identification of Low-Level Point Radioactive Sources Using a Sensor Network," ACM Transactions on Sensor Networks, Vol. 7, No. 3, pp. 493–504, September 2010.

12. S.S.Iyengar, Supratik Mukhopadhyay, Christopher Steinmuller, and Xin Li "Preventing Future Oil Spills with Software-Based Event Detection" IEEE Computers, Vol. 43, Issue 8, pp. 76–78, August 2010.

13. Kun Yan, Hsiao-Chun Wu, and S. Sitharama Iyengar, "Robustness Analysis and New Hybrid Algorithm of Wideband Source Localization for Acoustic Sensor Networks", IEEE Transactions on Wireless Communications, Vol. 9, Issue 6, pp. 2033–2043, June 2010.

14. Mengxia Zhu, Song Ding, Richard R. Brooks, Qishi Wu, Nageswara S.V. Rao, S. Sitharama Iyengar, "Fusion of Threshold Rules for Target Detection in Sensor Networks", ACM Transactions on Sensor Networks, Vol. 6 and Issue 2, pp. 1–7, February 2010.

15. Vasanth Iyer, S.S. Iyengar, R. Murthy, and M. B. Srinivas. "Computational Aspects of Sensor Network Protocols (distributed sensor network simulator)," Sensors & Transducers Journal, Vol. 6, Issue 1, pp. 69–91, July 2009.

16. Vasanth Iyer, S.S. Iyengar, G. Rama Murthy and M.B. Srinivas. "Distributed Source Coding for Sensor Data Model," International Journal of Simulation: Systems, Science & Technology (IJSSST), Vol. 10, No. 1, pp. 16–23, 2009.

17. Vasanth Iyer, S.S.Iyengar, G. Ramamurthy and M.B. Srinivas, "Computational Aspects of Sensor Network Protocols," Sensors & Transducers Journal, Vol. 5, Special Issue, pp. 69–91, March 2009.

18. Neeta Trivedi, S. Sitharama Iyengar, N. Balakrishnan, "Ripples: Message-Efficient, Coverage-Aware Clustering in Wireless Sensor and Actor Networks," International Journal of Communication and Distributed Systems, Vol. 2, Issue 1, pp. 112–134, January, 2009.

19. P.T. Krishna Kumar, V.V. Phoha, S.S. Iyengar, "Simulation of Robust Resonance Parameters Using Information Theory," Elsevier Annals of Nuclear Energy, Vol. 35, Issue 8, pp. 1515–1518, August 2008.

20. P.T. Krishna Kumar, V.V. Phoha, S.S. Iyengar, "Classification of Radio Elements Using Mutual Information: A Tool for Geological Mapping," Elsevier International Journal of Applied Earth Observation and Geo-information, Vol 10, Issue 3, pp. 305–311, September 2008.

21. Q. Wu, N.S.V. Rao, X. Du, S.S. Iyengar, V.K. Vaishnavi, "On Efficient Deployment of Sensors on Planar Grid," Elsevier's Computer Communications, Vol. 30, Issue 14–15, pp.2721–2734, October 2007.

22. S.S. Iyengar, Hsiao-Chun Wu, N. Balakrishnan, Shih Yu Chang, "Biologically Inspired Cooperative Routing for Wireless Mobile Sensor Networks," IEEE Systems Journal, Vol. 1, No. 1, pp. 29–37, Sep 2007.

23. R. Kalidindi, R. Kannan, S.S. Iyengar, A. Durresi. "Sub-Grid Based Key Vector Assignment: A Key Pre-Distribution Scheme for Distributed Sensor Networks,"

Journal of Pervasive Computing and Communications, Vol. 2, No. 1, pp. 35–45, March 2006.

24. Durresi, V. Paruchuri, R. Kannan, S.S.Iyengar, "Optimized Broadcast Protocol for Sensor Networks", IEEE Transactions on Computers, Vol. 54, No. 8, pp. 1013–1023, August 2005.

25. H.M.F. Aboelfotoh, S.S. Iyengar, and K. Chakrabarty, "Computing Reliability and Message Delay for Cooperative Wireless Distributed Sensor Networks Subject to Random Failures", IEEE Transactions on Reliability, Vol. 54, No. 1, pp. 145–155, March 2005.

26. Danyuang Zhang, Sibabrata Ray, Rajgopal Kanna, S. Sitharama Iyengar, "Subgroup Based Source Recovery Or Local Recovery for Reliable Multicasting", International Journal of Computer Applications, Vol. 12, No. 2, pp. 1–12, June 2005.

27. W. Ding, S.S. Iyengar, R. Kannan, W. Rummler, "Energy Equivalence Routing in Wireless Sensor Networks", Journal of Microcomputers and Applications, Vol. 28/8, Special Issue, pp. 467–475, August 2004.

28. R. Kannan, S.S. Iyengar, "Game-Theoretic Models for Reliable Path-Length and Energy-Constrained Routing with Data Aggregation in Wireless Sensor Networks", IEEE Transactions on Selected Areas of Communications, Vol. 22. No. 6, pp. 1141- 1150, August 2004.

29. R. Kannan, S.S. Iyengar, and S. Sarangi, "Sensor-Centric Energy-Constrained Reliable Query Routing for Wireless Sensor Networks", Journal of Parallel and Distributed Computing, Vol. 64, Issue 7, pp. 839–852, July 2004.

30. R.R. Brooks, M. Zhu, J. Lamb, S.S. Iyengar, "Aspect-Oriented Design of Sensor Networks", Journal of Parallel and Distributed Computing, Vol. 64, Issue 1, pp. 853–865, July 2004.

31. Q. Wu, S.S. Iyengar, N.S.V. Rao, J. Barhen, V. K. Vaishnavi, H. Qi, K. Chakrabarty, "On Computing the Route of a Mobile Agent for Data Fusion in a Distributed Sensor Network", IEEE Transactions on Knowledge and Data Engineering, Vol. 16, Issue 6, pp. 740–753, June 2004.

32. B. Krishnamachari, S.S. Iyengar, "Distributed Bayesian Algorithms for Fault-Tolerant Event Region Detection in Wireless Sensor Networks", IEEE Transactions on Computers, Vol 53, No. 3, March 1, 2004.

33. S. Sastry, S. S. Iyengar, N. Balakrishnan, "Sensor Technologies for Future Automation System" Journal of Sensor Processing Letters, Vol 2, pp. 1–9, 2004.

34. Qishi Wu, Nageswara S.V. Rao, Richard R. Brooks, S. Sithamara Iyengar, Mengxia Zhu, "On Computational and Networking Problems in Distributed Sensor Networks", Handbook on Sensor Networks, pp. 1–23, July 2004.

35. Rajgopal Kannan, Lydia Ray, Arjan Durresi, S. Sithamara Iyengar, "Security-Performance Trade- off of Inheritance Based Key Predistribution for Wireless Sensor Networks", Cornell University Library, pp. 1–23, 2004.

36. Guna Seetharaman, S.S. Iyengar, Ha V. Le, N.Balakrishnan, R. Logananthraj, "SmartSAM: A Multisensor Network Based Framework for Video Surveillance and Monitoring", Sensor Network Operations, pp. 631–647, John Wiley & Sons, Inc., September 2004.

37. R. Kannan, L. Ray, R. Kalidindi, S.S. Iyengar, "Threshold-Energy Constrained Protocol for Wireless Sensor Networks", Sensor Processing Letters, Vol. 1, No.1, pp. 79–85, December 2003.
38. S.S. Iyengar, S. Sastry, and N. Balakrishnan, "Foundations of Data Fusion for Automation", IEEE Instrumentation and Measurement Magazine, Vol. 6, Issue 4, pp 35–41, December 2003.
39. Kannan, S. Ray, S. Sarangi, S.S. Iyengar, "Minimal Sensor Integrity: Measuring Vulnerability in Sensor Deployments ", Information Processing Letters, Vol. 86, Issue 1, pp 49–55, April 2003.
40. Rajgopal Kanna, Lydia Ray, Ramaraju Kalidindi, and S. Sithamara Iyengar, "Threshold-Energy- Constrained Routing Protocol for Wireless Sensor Networks", Sensor Letters, Vol. 1, pp. 79–85, 2003.
41. K. Chakrabarty, S.S. Iyengar, H. Qi, and E.C. Cho, "Grid Coverage of Surveillance and Target Location in Distributed Sensor Networks", IEEE Transactions on Computers, Vol. 8, No. 3, pp. 1448–1453, Dec. 2002.
42. H. Qi, S.S. Iyengar and K. Chakrabarty, "Multi-Resolution Data Integration Using Mobile Agents in Distributed Sensor Networks", IEEE-Systems Man Cybernetics, Vol 31, No. 3, pp. 383–390, August 2001.
43. S.S. Iyengar and B. Jones, "Information Fusion Techniques for Pattern Analysis in Large Sensor Data Networks", Journal of Franklin Institute, Vol. 338, pp. 571–582, July 2001.
44. H. Qi, S.S. Iyengar, K. Chakrabarty, "Distributed Sensor Networks, A Review of Recent Research", Journal of Franklin Institute, Vol. 338, pp. 655–668, 2001.
45. S.S. Iyengar, K. Chakrabarty, H. Qi, " Introduction to the Special Issue on Distributed Sensor Networks for Real Time Systems with Adaptive Configurations", Journal of Franklin Institute, Vol 338, pp. 651–653. 2001.
46. R.R. Brooks, S.S.Iyengar, and S. Rai, "Comparison of Genetic Algorithm and Simulated Annealing for Cost Minimization in a Multi-Sensor System", Journal of Optical Eng. Vol. 37, Issue 2, pp. 505–516, February 1998.
47. Richard R. Brooks, S. Sitharama Iyengar, "Real-Time Distributed Sensor Fusion for Time-Critical Sensor Readings", Society of Photo Optical Instrumentation Engineers. Opt. Eng. Vol. 36, Issue 3, pp. 767 -779, March 1997.
48. J.R. Maheshkumar, V. Veeranna, S.S. Iyengar and R.R. Brooks, A New Computational Technique for Complementary Sensor Integration in Detection Localization Systems", Journal of Optical Engineering, Vol. 35, Issue 3, pp. 674–684, March 1996.
49. R. R. Brooks, S. S. Iyengar, J. Chen, "Automatic Correlation and Calibration of Noisy Sensor Readings Using Elite Genetic Algorithms", Artificial Intelligence, Vol. 84, 339–354, 1996.
50. Nageswara S. V. Rao, S. Sithamara Iyengar, "Distributed Decision Fusion Under Unknown Distributions", Journal of Optical Engineering, Vol. 35, Issue 3, pp. 617–624, March 1996.
51. S.S. Iyengar and R.L. Kashyap, "Introduction to the Special Issue on Parallel and Distributed Image and Sensor Signal Integration Problems", Journal of Franklin Institute, Vol. 5, pp. 16–32, 1995.

52. S. S. Iyengar and L. Prasad, "A General Computational Framework for Distributed Sensing and Fault-Tolerant Sensor Integration ", IEEE Transactions on Systems, Man, and Cybernetics, Vol. 25, No. 4, April 1995.
53. Sankar Krishnamurthy, S. Sitharama Iyengar, Ronald J. Holyer, and Matthew Lybanon, "Histogram- Based Morphological Edge Detector", IEEE Transaction on Geoscience and Remote Sensing, Vol. 32, No. 4, July 1994.
54. L. Prasad, S.S.Iyengar, R. Rao and R. L. Kashyap, "Fault-Tolerant Integration of Abstract Sensor Estimates Using Multi-Resolution Decomposition", Physical Review E, Vol. 49, No. 4, pp. 3452–3460, October 1993.
55. S. Sithamara Iyengar, Mohan B. Sharma, and R. L. Kashyap, "Information Routing and Reliability Issues in Distributed Sensor Networks", IEEE Transactions on Signal Processing, Vol. 40, No. 12, December 1992.
56. L. Prasad, S.S.Iyengar, R. L. Kashyap and R. N. Madan, "Functional Characterization of Fault Tolerant Integration in Distributed Sensor Networks", IEEE Transactions On Systems, Man, and Cybernetics, Vol.21, No.5, pp. 1082–1087, October 1991.
57. S. Sithamara Iyengar, R. L. Kashyap, and Rabinder N. Madan, "Distributed Sensor Networks - Introduction to the Special Section", IEEE Transactions on Systems, Man, and Cybernetics, Vol. 21, No. 5. September/October 1991.
58. S. Gulati, S.S. Iyengar, and Barhen, "The Pebble Crunching Model for Fault Tolerant Load Balancing in Hypercube Ensembles", Computer Journal, Vol. 33, No. 3, June 1990.

Refereed Conference Papers and Book Chapters

1. Thejas G.S., Kianoosh G. Boroojeni, Kshitij Chandna, Isha Bhatia, S.S. Iyengar, and N.R. Sunitha. Deep Learning-based Model to Fight Against Ad Click Fraud. In ACM Southeast Conference (ACMSE 2019), 2019, pp. 176–181.
2. Thejas G.S., Jayesh Soni, Kshitij Chandna, S. S. Iyengar, N. R. Sunitha, and Nagarajan Prabakar. Learning-Based Model to Fight against Fake Like Clicks on Instagram Posts. In IEEE SoutheastCon, 2019, pp. 1–8. In press.
3. Thejas G.S., T. C. Pramod, S. S. Iyengar, and N. R. Sunitha. 2018. Intelligent Access Control: A Self-Adaptable Trust-Based Access Control (SATBAC) Framework Using Game Theory Strategy. In Proceedings of International Symposium on Sensor Networks, Systems and Security, Nageswara S.V. Rao, Richard R. Brooks, and Chase Q. Wu (Eds.). Springer International Publishing, Cham, pp. 97–111.
4. S. S. Iyengar, Kianoosh G. Boroojeni, N. Balakrishnan, " Introduction to Distributed Sensor Networks", Springer, 2014.
5. S. S. Iyengar, Kianoosh G. Boroojeni, N. Balakrishnan, " Expectation-Maximization for Acoustic Source Localization", Springer, 2014.

6. S. S. Iyengar, Kianoosh G. Boroojeni, N. Balakrishnan, " Coordinate-Free Coverage in Sensor Networks via Homology", Springer, 2014.

7. S. S. Iyengar, Kianoosh G. Boroojeni, N. Balakrishnan, " Coverage Assessment and Target Tracking in 3D Domains", Springer, 2014.

8. S. S. Iyengar, Kianoosh G. Boroojeni, N. Balakrishnan, " A Stochastic Preserving Scheme of Location Privacy", Springer, 2014.

9. S. S. Iyengar, Kianoosh G. Boroojeni, N. Balakrishnan, " Region-Guarding in 3D Areas", Springer, 2014.

10. EG Prathima, T Shiv Prakash, KR Venugopal, SS Iyengar, LM Patnaik, "SDAMQ: Secure Data Aggregation for Multiple Queries in Wireless Sensor Networks," Procedia Computer Science, Vol. 89, pp. 283–292, 2016.

11. Raghavendra, S., et al. "Index Generation and Secure Multi-user Access Control over an Encrypted Cloud Data." Procedia Computer Science, Vol. 89, pp. 293–300, 2016.

12. Mingming Guo, Niki Pissinou, and S.S. Iyengar, "Privacy-Aware Mobile Sensing in Vehicular Networks", International Conference on Computing, Networking and Communication (ICNC'16)

13. Mingming Guo, Niki Pissinou, S. Sitharama Iyengar, "Pseudonym-based anonymity zone generation for mobile service with strong adversary model". CCNC 2015: 335–340.

14. Samia Tasnim, Mohammad Ataur Rahman Chowdhury, Kishwar Ahmed, Niki Pissinou, S. Sitharama Iyengar, "Location aware code offloading on mobile cloud with QoS constraint". CCNC 2014: 74–79.

15. CR Yamuna Devi, B Shivaraj, SH Manjula, KR Venugopal, SS Iyengar, LM Patnaik, "Multi-hop optimal position based opportunistic routing for wireless sensor networks," IEEE Region 10 Symposium, pp 121–125, 2014.

16. T Shiva Prakash, KB Raja, KR Venugopal, SS Iyengar, LM Patnaik, "Base Station Controlled Adaptive Clustering for QoS in Wireless Sensor Networks," International Journal of Computer Science and Network Security, Vol. 14, Issue 2, 2014.

17. T Shiva Prakash, KB Raja, KR Venugopal, SS Iyengar, LM Patnaik, "Fault Tolerant QoS Adaptive Clustering for Wireless Sensor Networks," Proceedings of Ninth International Conference on Wireless Communication and Sensor Networks, pp167–175, 2014

18. Vasanth Iyer, S. Sitharama Iyengar, Niki Pissinou, and Shaolei Ren, "SPOTLESS: Similarity Patterns Of Trajectories in Label-lEss Sensor Streams", the 5th International Workshop on Information Quality and Quality of Service for Pervasive Computing 2013, San Diego, CA.

19. Prakash T Shiva, Kiran B Raja, KR Venugopal, SS Iyengar, Lalit M Patnaik, "Link-reliability based two-hop routing for QoS guarantee in Wireless Sensor Networks," Wireless Personal Multimedia Communications (WPMC), 2013 16th International Symposium on, pp 1–6, 2013

20. Sivasankari H, Leelavathi R, Shaila K, Venugopal K R, S. S. Iyengar, Patnaik L. M, "Energy Efficient Adaptive Cooperative Routing with Multiple

Sinks in Wireless Sensor Networks", In 8Th IEEE Conference on Industrial Electronics and Applications (ICIEA 2013) in Melbourne, Australia.

21. T Shiva Prakash, Kiran B Raja, KR Venugopal, SS Iyengar, Lalit M Patnaik, "Traffic-differentiated two-hop routing for QoS in wireless sensor networks," Cyber-Enabled Distributed Computing and Knowledge Discovery (CyberC), 2013 International Conference on, pp 356–363, 2013.

22. R Tanuja, MK Rekha, SH Manjula, KR Venugopal, SS Iyengar, LM Patnaik, "Elimination of black hole and false data injection attacks in wireless sensor networks," Proceedings of the Third International Conference on Trends in Information, Telecommunication and Computing, pp 475–482, 2013.

23. H Sivsankari, R Leelavathi, K Shaila, KR Venugopal, SS Iyengar, LM Patnaik, "Energy Efficient Adaptive Cooperative Routing (EEACR) with multiple sinks in Wireless Sensor Networks," Industrial Electronics and Applications (ICIEA), 2012 7th IEEE Conference on, pp 676–681, 2012.

24. K.R. Venugopal and L.M. Patnaik, S. S. Iyengar, "Secure Reputation Update for Target Localization in Wireless Sensor Networks", ICIP 2012, CCIS 292, pp. 109–118, Springer-Verlag Berlin Heidelberg 2012.

25. H. Sivasankari, Aparna R, Venugopal Kr, S S Iyengar and L M Patnaik, "TGAR: Trust Dependent Greedy Anti-Void Routing in Wireless Sensor Networks (WSNS)", Proceedings of LNICEE Conference on ITC, August, 2012.

26. Sandeep Khurana, Nathan Brener, Werner Benger, Somnath Roy, Sumanta Acharya, Marcel Ritter,

27. Vasanth Iyer, S. Sitharama Iyengar. MODELING UNRELIABLE DATA AND SENSORS: Using F- measure Attribute Performance with Test Samples from Low-cost Sensors, in ICDMW 2011, IEEE International Conference on Data Mining-Workshops (ICDMW 11), 2011 Vancouver, Canada.

28. Vasanth Iyer, S. Sitharama Iyengar, N. Parameswaran, Garmiela Rama Murthy and Mandalika B. Srinivas, Machine Learning and Data mining Algorithms for Predicting Accidental Forest Fires, In Proc. International Conference on Sensor Technologies and Applications SENSORCOMM, 17–21 August. 2011.

29. Jong-Hoon Kim, Gokarna Sharma, Noureddine Boudriga, and S. Sitharama Iyengar, "SPAMMS: A Sensor-Based Pipeline Autonomous Monitoring and Maintenance System", 2010 Second International Conference on Communication Systems and Networks (COMSNETS). January 2010.

30. Vasanth Iyer, S.S.Iyengar, G. Ramamoorthy, Kannan Srinathan, Rakee and M.B. Srinivas, Intelligent Networks Sensor Processing of Information using Key Management. In Proc. 4th International Conference on Sensing Technology - ICST, Lecce, Italy, 2010.

31. Vasanth Iyer, S.S. Iyengar, G. Rama Murthy, Kannan Srinathan, Vir Phoha, and M.B. Srinivas, "INSPIRE-DB: Intelligent Networks Sensor Processing of Information using Resilient Encoded-Hash DataBase", In Proc. Fourth International Conference on Sensor Technologies and Applications SEN-SORCOMM 2010, Venice, Italy.

32. P.T.Krishna Kumar, Suhas Madhusudhana, P.T. Vinod, S. Sitharama Iyengar "Mitigation of Toxicity in Marine Mussels by Autonomous Mobile Agents", IEEE 2010 - International Conference on Wireless Communication and Sensor Computing, January 2010.

33. Vasanth Iyer, S. S. Iyengar, N. Balakrishnan, V. Phoha, and M. B. Srinivas. Farms: Fusionable Ambient Renewable MACS. In Proc. IEEE Sensors Applications Symposium SAS, pages 169–174, 17–19 Feb. 2009. Doi: 10.1109/SAS. 2009.4801800.

34. Vasanth Iyer, S. Sitharama Iyengar, N. Balakrishnan, Vir. Phoha and G. Rama Murthy, "Distributed Source Coding for Sensor Data Model and Estimation of Cluster Head Errors Using Bayesian and K-Near Neighborhood Classifiers in Deployment of Dense Wireless Sensor Networks ", SENSORCOMM, Athens/Vouliagmeni, Greece, June 18–23, 2009.

35. Vasanth Iyer, S.S.Iyengar, G. Ramamurthy and M.B. Srinivas, "Multi-Hop Scheduling and Local Data Link Aggregation Dependent QoS in Modeling and Simulation of Power-Aware Wireless Sensor Net- works", International Wireless Communications & Mobile Computing, Leipzig, Germany, June 21–24 2009.

36. Suman Kumar, S. S. Iyengar, Ravi Lochan, Urban Wiggins, Kanwalbir Sekhon, Promita Chakraborty, and Raven Dora, "Application of Sensor Networks for Monitoring of Rice Plants: A Case Study. IRADSN 2009. Hong Kong, China May 2009."

37. Srivathsan Srinivasagopalan, Costas Busch, and S. Sitharama Iyengar, "Brief Announcement: Universal Data Aggregation Trees for Sensor Networks in Low Doubling Metrics",: Algosensors 2009, LNCS 5804, pp. 151–152, 2009

38. Q. Wu, M. Zhu, N. S. V. Rao, S. S. Iyengar, R. R. Brooks, M. Meng, "An Integrated Intelligent Decision Support System based on Sensor and Computer Networks," Systems of Systems Engineering, M. Jamshidi (Editor) 2008.

39. Shuanging Wei, Rajgopal Kannan, S.S. Iyengar and Nageswara S. Rao, "Energy Efficient Estimation of Gaussian Sources Over Inhomogeneous Gaussian Mac Channels", IEEE GLOBECOM Conference 2008, 30 November- 4 December 2008.

40. Engchun Cho, Srivatsan Srinivasagopalan, N. Balakrishnan, S.S. Iyengar, "Distributed Sensor Network Deployed on Eisenstein Grids", The IASTED International Symposium on Distributed Sensor Network (Dsn'08), November 16–18, 2008, Orlando, Florida. 2008.

41. Hsiao-Chun Wu, Kun Yan, S.S. Iyengar, "Robustness Analysis of Source Localization Using Gaussianity Measure", IEEE GLOBECOM 2008, 30 November- 4 December, 2008.

42. N. Rao, M. Shankar, Jren-Chit Chin, David Yau, S. Srivathsan, S.S.Iyengar, Y. Yang, J. Hou, "Identification of Low -Level Point Radiation Sources Using a Sensor Network", Proceedings of International Conference on Information Processing in Sensor Networks, April 22–24,2008, St. Louis, Missouri. 2008.

43. Suman Kumar, Srivatsan Srinivasagopalan, Seung-Jong Park, S.S. Iyengar, "Estimating Data Redundancy in Sensor Networks", Third International

Innovations and Real-Time Applications of Distributed Sensor Networks Symposium, Shreveport, Louisiana, November 26–27 2007.

44. Srivatsan Srinivasagopalan, S.S. Iyengar, "Minimizing Latency in Wireless Sensor Networks - A Sur- vey", Third IASTED Conference on Advances in Computer Science and Technology, Phuket, Thailand, April 2–4, 2007.

45. N. S. V. Rao, S. Ding, S. S. Iyengar, "Difference Triangulation Method under Monotone Function of Distances," Ninth ONR/GTRI Workshop on Target Tracking and Sensor Fusion, abstract, 2006.

46. M. Zhu, R.R. Brooks, Q. Wu, N. S. V. Ramo, S. Ding, S. S. Iyengar, "Fusion of Threshold Rules for Target Tracking in Self-Organizing Sensor Networks," Ninth ONR/GTRI Workshop on Target Tracking and Sensor Fusion, abstract, 2006.

47. S. Srivathsan and S.S. Iyengar, "Reliability in Wireless Sensor Networks", Proceedings of the Second IEEE International Workshop on Next Generation Wireless Networks. December 18–21 2006.

48. Tom Rishel, A. Louise Perkins, Sumanth Yenduri, Farnaz Zand, "Augmentation of a Term/Documentation Matrix with Part-of-Speech Tags to Improve Accuracy of Latent Semantic Analysis", Proceedings of the 5Th WSEAS International Conference on Applied Computer Science (pp. 573–578), Hangzhou, China, April 16–18, 2006.

49. Sumanth Yenduri, Kanthi Kumar Adapa, S.S. Iyengar, Ravi Paruchuri, "A Methodology to Increase Security in Wireless Networks", Proceedings of the 8th World Multi-Conference on Systemics, Cybernetics and Informatics, SCI 2004, - Orlando, Florida, USA. July 18–21, 2004.

50. R. Kalidindi, V. Parachuri, S. Basavaraju, C. Mallanda, A. Kulshrestha, L. Ray, R. Kannan, A. Durresi, and S.S. Iyengar, "Sub-Grid Based Key Vector Assignment: A Key Pre-Distribution Scheme for Distributed Sensor Networks", Proceedings of the 2004 International Conference on Wireless Networks (Icwn'04), June 21–24, 2004, Las Vegas, Nevada, USA.

51. C. Mallanda, S. Basavaraju, A. Kulshrestha, R. Kannan and S.S.Iyengar, " Secure Cluster Based Energy Aware Routing for Wireless Sensor Networks ", the 2004 International Conference on Wireless Networks (Icwn'04), Las Vegas, Nevada, USA. June 21–24, 2004.

52. Guna Seetharaman, Ha V. Le. S.S.Iyengar, R. Logananthraj, "SmartSAM: A Multisensor Network Based Framework for Video Surveillance and Monitoring", Sixteenth International Symposium on Mathematical Theory of Networks and Systems, Belgium July 5–9, 2004.

53. Eungchun Cho, S.S. Iyengar, "Application of Eisenstein Integers for Efficient Placement of Sensors in a Distributed Sensor Network", First International Workshop on Algorithmic Aspects of Wireless Sensor Networks, Turku, Finland, July 16 2004.

54. Arjan Durresi, Vamsi Paruchuri, Rajgopal Kannan, S.S. Iyengar," A Lightweight Protocol for Data Integrity in Sensor Networks", IEEE 2004. ISSNIP 2004.

55. S. S. Iyengar, G. Seetharaman, R. Kannan, A. Durresi, S. Park, B. Krishna-machari, R. R. Brooks and J. Morrison," Next Generation Distributed Sensor Networks", Proceedings of office of Naval Research, September 5–6, 2004, USA.

56. R. Kannan, Lydia Ray, S.S. Iyengar and R. Kalidindi, "Max-Min Length-Energy-Constrained Routing in Wireless Sensor Networks ", 1St European Conference Workshop on Wireless Sensor Networks, Berlin, Germany, January 18–21 2004.

57. Vamsi Paruchi, Shivakumar Basavaraju, Arjan Durresi, Rajgopal Kannan, and S. S. Iyengar, "Random Asynchronous Wakeup Protocol for Sensor Networks", Proceedings of the First International Conference on Broadband Networks. 2004.

58. Elias G. Khalaf, S. Sithamara Iyengar, "Scalable Reliable Multicast Using Receiver Grouping, "Proceedings of the International Conference on Internet Computing, Las Vegas, 2004.

59. R. Kalindindi, R. Kannan, S.S. Iyengar and L. Ray, "An Energy Efficient Mac Protocol for Sensor Networks", International Workshop on Wireless Networks, Las Vegas, NV, July 03.

60. R.R. Brooks, Matthew Pirretti, Mengxia Zhu, S.S. Iyengar, "Adaptive Routing Using Emergent Protocols in Wireless Ad Hoc Sensor Networks", Proceedings of SPIE Conference, 6–8 August, Vol. 5205, 2003.

61. R. Kannan, S.S. Iyengar and V. Kumar, "A New Framework for Quantifiable Data Security in Sensor Networks", 17th Annual IFIP WG 11.3 Working Conference on Data and Applications Security, Estes Park, CO, August 2003.

62. R. Kalindindi, R. Kannan, S.S. Iyengar and L. Ray, "Distributed Energy Aware Mac Layer Protocol for Wireless Sensor Networks", International Workshop on Wireless Networks, Las Vegas, NV, July 2003.

63. R. Kannan, R. Kalidindi, S. S. Iyengar, V. Kumar, " Energy and Rate Based Mac Protocol for Wireless Sensor Networks ", Proceedings of the ACM Special Interest Group on Management of Data (SIGMOD) Record, Vol. 32, No. 4, December, 2003.

64. R. R. Brooks, M. Pirretti, M. Zhu, S. S. Iyengar, "Distributed Adaptation Methods for Wireless Sensor Networks", Proceedings of Globe Com 2003 Conference, San Francisco, CA, Dec 2–5, 2003.

65. B. Krishnamachari, S.S. Iyengar, " Efficient and Fault-Tolerant Feature Extraction in Wireless Sensor Networks ", Proceedings of Information Processing in Sensor Networks, Palo Alto, CA, pp. 488–501, April 2003.

66. R. Kannan, S. Sarangi, S.S. Iyengar and L. Ray, "Sensor-Centric Quality of Routing in Sensor Net- works", Proceedings of IEEE Computer & Communications INFOCOM, Volume 64, No. 7, pp. 839–852, April 2003.

67. Bhaskar Krishnamachari, and Sithamara Iyengar, "Self-Organized Fault-Tolerant Feature Extraction in a Distributed Wireless Sensor Network", Proceedings of Information Processing in Sensor Networks, Palo Alto, CA, April 2003.

68. R. Kannan, S. Sarangi, S.S. Iyengar, " A Simple Model for Reliable Query Reporting in Sensor Networks ", Fifth International Conference on Information Fusion, pp. 754–759, Annapolis, MD, July 2002.

69. R. Kannan, S. Ray, S. Sarangi, S.S. Iyengar, "Minimal Sensor Integrity: Measuring Integrity in Sensor Deployments", International Conference on Parallel Processing (ICPP), Vancouver, British Columbia, July, 2002.

70. S.S. Dhillon, K. Chakrabarty, and S.S. Iyengar, "Sensor Placement for Grid Coverage under Imprecise Detections", Proceedings of the International Conference on Information Fusion (Fusion 2002), pp. 1581–1587, 2002.

71. Xiaoling Wang, Halrong Qi, S. Sithamara Iyengar, "Collaborative Multi-Modality Target Classification in Distributed Sensor Networks", Proceedings of the Fifth International Conference on Information Fusion. Knoxville, Tennessee. 2002.

72. S. Sithamara Iyengar, Qishi Wu, "Computational Aspects of Distributed Sensor Networks", Proceedings of the International Symposium on Parallel Architectures, Algorithms and Network. 2002.

73. Rajgopal Kannan, Sudipta Sarangi, Sibabrata Ray, S. S. Iyengar, "Minimal Sensor Integrity in Sensor Grids", Proceedings of the International Conference on Parallel Processing 2002.

74. Krishnendu Chakrabarty, S. S. Iyengar, "Sensor Placement in Distributed Sensor Networks Using a Coding Theory Framework", IEEE - ISIT, Washington, Dc, June 24–29, 2001.

75. H. Qi, X. Wang, S.S.Iyengar, K. Chakrabarty, "Multi Data Fusion in Distributed Sensor Networks Using Mobile Agents", Proceedings of 4Th Annual Conference on Information Fusion, Vol 1, Fusion 2001, Montreal, Quebec, Canada, 7–10 August, 2001.

76. K. Chakrabarty, S.S.Iyengar, H. Qi and E.C. Cho," Coding Theory Framework for Target Locations in Distributed Sensor Networks ", Proceedings of International Symposium on Information Technology: Coding and Computing, Las Vegas, Nevada, April 2001.

77. H. Qi, S.S.Iyengar, and K. Chakrabarty, "Distributed Multi-Resolution Data Integration Using Mobile Agents", Proceedings of IEEE Aerospace Conference, March 2001.

78. Sumeet Dua and S.S.Iyengar, "Detection of Frequent Episodes in Web Access Logs and Dynamic Web Server: A Case for a Unified Framework", Proceedings of IEEE-Asset, (March 2000), Dallas, Texas.

79. S.S.Iyengar, "A Road Map to Information Technology for the 21St Century", Proceedings of the 14Th Institution of Engineers, Hyderabad, India. Dec 17–19, 1999.

80. S.S.Iyengar and B. Jones "Information Fusion in Manufacturing Environment", Proceedings of International Conference on Information Technology Integration for Manufacturing. Bangalore, India Dec.28–30, 1998.

81. J. Zachary and S.S.Iyengar, "Three Dimensional Data Fusion for Biomedical Surface Reconstruction ", Proc. SPIE Aerosense 97 Sensor Fusion: Architectures, Algorithms, and Applications, April 1997.

82. R.R. Brooks and S.S.Iyengar, "Minimizing Cost of Redundant Sensor Systems with Non-Monotone and Monotone Search Algorithms ", Proceedings of 1997 IEEE Reliability and Maintainability Symposium, Philadelphia, January 1997.

83. R. R. Brooks, S. S. Iyengar, N. S. V. Rao, "Sensor Fusion Survey: Sensors, Statistics, Signal Processing and Neural Networks," Third International Conference on Neural Networks and Their Applications (NEURAP'97), Marseille, France, March 1997.

84. R.R. Brooks and S.S.Iyengar, "Maximizing Multi-Sensor System Dependability", Proceedings of IEEE Conference on Multi-Sensor Fusion and Integration for Intelligent Systems, Washington, D.C., December 1996.

85. R.R. Brooks and S.S.Iyengar, "Dynamic Sensor Fusion", Proceedings of the Workshop on Foundations of Information/Decision Fusion: Applications to Engineering Problems, Aug. 7–9, Washington D.C. 1996.

86. R.R. Brooks and S.S.Iyengar, "Methods of Approximate Agreement for Multisensor Fusion", SPIE Proceedings Signal Processing, Sensor Fusion and Target Recognition Iv, Orlando, Fl., April 1995.

87. R.R. Brooks, S.S.Iyengar, and J. Chen, "Self Calibration of a Noisy Multiple Sensor System with Genetic Algorithms," Proceedings of The SPIE'S Conference on Intelligent Systems and Manufacturing, October 1995.pp.20–30, Vol. 25–89, Philadelphia, Pennsylvania 1995.

88. R.R. Brooks, and S.S.Iyengar, "Optimal Matching Algorithm for Multidimensional Sensor Readings", Proceedings of the SPIE'S Conference on Intelligent Systems and Manufacturing, October 1995. SPIE Volume 2589, pp. 91–99, Philadelphia, Pennsylvania. October 1995.

89. R.R. Brooks, S.S.Iyengar, "Robot Algorithm Evaluation by Simulating Sensor Faults," Signal Processing, Sensor Fusion, Target Recognition Iv, EDS. Kadar and Libby, SPIE, Bellingham, West Virginia, Proceedings of SPIE International Symposium on Aerospace/Defense Sensing, Dual Use Photonics, Orlando, Florida, and April 1995.

90. D. Nadig, S.S.Iyengar, and D. N. Jayashima, "A Versatile Architecture for Distributed Sensor Integration", Proceedings of IEEE-South Conference, March 1993.

91. S. Trivedi, B. Jones and S.S.Iyengar, "Reconstruction of Possible Systems with Incomplete Information", Proceedings of 32 Southeast ACM Conference, March 1994.

92. L. Prasad, S.S. Iyengar, R. Rao, "Fault-Tolerant Integration of Abstract Sensor Estimates Using Multiresolution Decomposition", Systems, Man and Cybernetics, 1993. Conference Proceedings of the International Conference on 'Systems Engineering in the Service of Humans'. October 1993.

93. D. Nadig and S. S. Iyengar, "A New Architecture for Distributed Sensor Integration", Proceedings of IEEE Southeast Conference. April 1993.

94. L. Prasad, L., S.S.Iyengar, R. Rao, and R.L. Kashyap, "Fault-Tolerant Integration of Abstract Sensor Estimation Using Multi-Resolution Decomposition," Proceedings of IEEE-SMC on Distributed Sensor Networks and Data Fusion, Paris, France, Oct 1993.

95. P. Graham and S.S.Iyengar, "Double and Triple Step Incremental Linear Interpolation", Proceedings of the 1993 Symposium on Applied Computing, Feb 1993.

96. R. Brooks, and S.S.Iyengar, "Algorithms for Resolving Inter-Dimensional Consistencies in 'Redundant Sensor' Arrays", Proceedings of Indo-US Workshop on Parallel and Distributed Signal and Image Integration Problems, Dec 1993.

97. S.S.Iyengar, "Distributed Sensing and Fault Tolerant Sensor Integration", Proceedings of IEEE-Southcon '92 Conference, March 10–12, 1992.

98. L. Prasad, S.S.Iyengar, R. L. Kashyap and R. N. Madan, "Functional Characterization of Sensor Integration in Distributed Sensor Networks", Proceeding Fifth International Parallel Processing Symposium, April - May 1991.

99. R. L. Kashyap, S.S.Iyengar, and R. N. Madan, "A Tree Architecture for Sensor Fusion Problems", Proc. of SPIE'S Technical Symposium on Sensor Fusion, Orlando, Florida April 1990.

100. D. Thomas and S.S.Iyengar, "A Distributed Sensor Network Structure with Fault - Tolerant Facilities," Proc. Of the 89 SPIE'S Symposium on Advances in Intelligent Systems, Philadelphia, Pennsylvania, Nov. 1989.

101. Vasanth Iyer, S.S. Iyengar, Niki Pissinou. "Using Event Log Performance and F-Measure Attribute Selection". Book-Titled "Intelligent Sensor Networks: The Integration of Sensor Networks", Signal Processing and Machine Learning. ISBN: 978-1-43-989281-7 published by Taylor & Francis, December 12, 2012. pp 32–52, 2012.

102. H Sivasankari, R Leelavathi, K Shaila, KR Venugopal, SS Iyengar, LM Patnaik, "Dynamic Cooperative Routing (DCR) in Wireless Sensor Networks," Advances in Communication, Network, and Computing, Springer Verlag, pp 87–92, 2012.

103. M. Zhu, R. R. Brooks, S. Finh, Q. Wu, N. S. V. Rao, S. S. Iyengar, "Chebyshev's Inequality-based Multi-Sensor Data Fusion in Self-Organizing Sensor Networks," 2011, Chapman and Hall Publications.

104. S. Srivathsan, N. Balakrishnan, S.S. Iyengar, "Scalability in Wireless Mesh Networks", In "Handbook of Wireless Mesh Networks". Publication by Springer (London) - 2008.

105. S. Srivathsan, N. Balakrishnan, S.S. Iyengar, "Critical Feature Detection in Cockpits - Application of AI in Sensor Networks", Published in "Computational Intelligence in Multimedia Processing: Recent Advances". Springer-Verlag - Mar 28, 2008, ISBN-13: 9783540768265, 400pp.

106. Q. Wu, N.S.V. Rao, R. R. Brooks, S.S Iyengar, M. Zhu, "On Computational and Networking Problems in Distributed Sensor Networks", In Handbook of Sensor Networks: Compact Wireless and Wired Sensing Systems, Mohammad Ilyas (Editor), CRC Press LLC, August 2004.

107. S.S.Iyengar and S. B. Mohan, "Information Routing in Distributed Sensor Networks - Special Analysis in One Or Two Dimensions", Editor: S. Prasad and R. L. Kashyap, Oxford IBH Publishing Co., Pvt. Ltd.

Index

© Springer Nature Switzerland AG 2020
P. Sniatala et al., *Fundamentals of Brooks–Iyengar Distributed Sensing Algorithm,*
https://doi.org/10.1007/978-3-030-33132-0

Printed in the United States
By Bookmasters